WHO CUT GOD'S HAIR

*A Sensible Disquisition Into God, Religion,
And Other Weighty Matters*

GARRETT GLASS

Who Cut God's Hair

A Sensible Disquisition Into God, Religion,
And Other Weighty Matters

By Garrett Glass – Author

Website: inmankindsimage.com

Cover Image: Michelangelo, frescoes on the ceiling of The Sistine Chapel, Vatican
Cover Design: Garrett Glass and Pamela Trush
Layout: Pamela Trush of delaney-designs.com

Copyright © 2013
First Printing 2013 Revised Edition 2017

ISBN 13 Digit: 978-0-9911106-3-6
ISBN 10 Digit: 0-9911106-3-3

Printed in the United States of America

Dedicated to Martha Whitehead —

In appreciation for the cheerful encouragement she has given me as a writer.

TABLE OF CONTENTS

PART THREE - FAITH AND ATHEISM

ACKNOWLEDGMENTS

This book began as a personal exercise to determine more precisely what it was I believed about God. Readers of my two novels on early Christianity, *Jehoshua: Signs and Wonders*, and *Jehoshua: Conflagration,* understood I had written these fictional histories without the miracles one finds in Biblical accounts of Jesus and the earliest Christians. Readers interpreted that to mean I was an atheist.

I am happy to describe myself as an atheist, but the term *atheist* conveys one very particular meaning in modern life: the atheist does not believe God exists. Now I certainly don't believe God exists in a material sense, "out there in the real world." But even though like any atheist, I insist that God is a creation of our mind, and I have no desire to worship such a God, I have always felt that there was something of substance about this God of the mind. I could never quite figure out what it was that struck me as true.

In 2013, I decided to put my thoughts down on paper for my benefit, to see what it was about God that appeared both imaginary and real. I was determined to do no research for this effort, because once I started down that path, I would find myself writing about what other people thought about God. Besides, so much has been written on this subject, that I might never be able to finish the research, knowing how one topic of investigation leads to another.

So I forced myself to think through my beliefs, organize them, probe them for falsehoods or faulty logic, and write them down. I self-published the result, *Who Cut God's Hair*, in 2013. At the time, the book satisfied my deep curiosity about God, and I understand it did so for many of my readers. It is virtually impossible to say or write anything new about God – he is one of the most discussed characters in human history. My best hope for *Who Cut God's Hair* was to express ideas somewhat differently, and arrange them in a novel way that would shed some interesting light on God and why people believe in him.

One of the beautiful things about writing a book about religion and philosophy – especially when the topic is God – is that readers like to explore these topics in great detail. As I began to accept invitations from book clubs and churches to discuss the ideas in the book, more ideas about God surfaced that I thought deserved to be in the book. A real breakthrough came when I was invited by some of my oldest friends, David and Nancy Napalo, to speak at a Sunday service for the North Shore Unitarian Church, near Chicago. This is a congregation of Unitarian Universalists, and the UU's, as they call themselves, are different from other religious congregations. You will find Buddhists, Christians, Muslims, Jews, Hindus, and quite a number of atheists/agnostics at UU services.

I had to find a way to make my comments applicable to such a diverse group of believers and non-believers. My talk veered away from the ideas covered in *Who Cut God's Hair*, so that I could address the important commonalities which exist between believers and non-believers alike. Matters proceeded further along these lines when Alane Callander – a friend

I have known from grade school – invited me to speak at her congregation, the Unitarian Universalist Fellowship of Fredericksburg, VA. By now, I was becoming very familiar with UU congregations, and I needed to be equipped with some answers to a frequent question: atheists assert God is a creation of our mind, but is there any scientific evidence to back up this assertion? This was the point where I had to begin serious research.

I discovered that indeed there was a growing body of evidence that the experiences we all have with God are internally generated, in our mind. We live at a time of exciting discoveries regarding God-belief and religion, and these discoveries are being made by anthropologists, neuroscientists, evolutionary biologists, psychologists, social scientists, and other specialists. I began discussing these findings with other UU congregations, such as the Unitarian Universalist Fellowship of Winston-Salem, NC. By the time I had a meeting with Eric Townson and his fellow humanists at Humanism with Heart, also in Winston-Salem, I realized *Who Cut God's Hair* simply had to be rewritten with all the new ideas that were fomenting in my mind.

Thus began the hard work of putting fingers to keyboard, and rewriting over 90% of *Who Cut God's Hair*. It was time to call to aid some of my closest friends – those stalwarts who have the time and interest in proofreading drafts of the new edition of *Who Cut God's Hair*. Proofreading is hard work, and it involves not just spotting misspellings and grammatical errors, but detecting inconsistencies in the text, and punching the arguments about time and again to make sure they are defended as best as possible. We self-published authors have to impose on our good friends for providing these important services. A newcomer to my group of supporters, Penny Powers, helped me reshape the structure of the book, by pointing out where the narrative lacked direction, and how topics could be organized to provide cohesion to the overall message. And then there were those who have helped me on my previous books, who assisted me once again, and without whom I would have been left with a very inferior product. My appreciation goes out to Bob Trezevant, Martha Whitehead, Sy and Daiva Banaitis, and Alane Callander for their ongoing, vital support.

Finally, there were those who put up with me when I was travelling, allowing me to spend time in their home, where I sequestered myself in a room with my laptop, and appeared to do my best to ignore them. I owe a debt of gratitude to Beth and Windell Murphy, Lynne and Rick Palmore, Eshagh and Rose Shaoul, and Debbie and Michael Mitchell, who gave me the gifts of comfort and isolation to write the final chapters of *Who Cut God's Hair*.

With all these benefactors, new and old, I am determined to maintain my friendship. Therefore, I am obliged to say here and now, and unequivocally, that all errors in this book are my responsibility alone, as are all the arguments presented here. As an author, I have the obligation to do the final checking for errors, and create a product that ultimately reflects my views, and not necessarily those of anyone else.

I do not have specialty training in the scientific fields discussed in the opening chapters of this book. I could best be described as a dilettante in areas such as psychology, neuroscience, anthropology, and so on. Back in the eighteenth century, a dilettante was a respectable person, having an informed interest in a variety of subjects, the better to make sense of the world. Today the word has a pejorative meaning, since we live in an age of specialization, where technical knowledge is highly prized. The specialist has a professional reputation to defend and does not often stray into other fields in which they are not trained.

If I were a specialist in any of the scientific fields discussed here, I probably would have never written this book. As a generalist – a dilettante – I have the luxury of looking across all these different disciplines and finding commonalities and some meaning to the questions surrounding God-belief. You'll find the culmination of my dilettantism in Chapter Eight of this book, in which I present my personal model describing how we are susceptible to developing a belief in God from our infancy to our pre-adult years.

This model, with five different stages corresponding to important milestones in child development, is entirely speculative in nature. Parts of the model are backed up by current scientific research in different fields – and as I have said, I've tried to represent this research as accurately as possible. Specialists in any of these different fields would have difficulty endorsing this model, because it is multidisciplinary and ventures into fields in which they do not have the same technical expertise. Moreover, the model mixes hard science discoveries in fields like neuroscience and biology, with soft science disciplines such as psychology and comparative mythology. The model therefore has to be judged on different standards. Does it provide a cohesive and comprehensive picture for how God-belief might develop? Is this picture at least plausible, if not likely? Does the model provide any new insights into the way in which God-belief arises in most people? I have included my God-belief model in this book because I think it passes these standards.

For those contemporary authors I have cited in this book – scientists, professionals and scholars such as Andrew Newberg, Mark Waldman, John Wathey, and Chris Minnick – I have done my best to quote them accurately and present their views faithfully. Their important contributions to a fuller and deeper understanding of God-belief require no less, and to the extent I have not represented their views faithfully, the errors are entirely mine.

Most writers discover that the task of putting thoughts on paper is all-consuming, lonely work. My wife has learned not to interrupt me when I get that vacant, expressionless glaze over my eyes. She knows I am thinking deep thoughts and do not want to lose the train of that internal conversation. This is true, though my thoughts are never as deep as I believe. I have also told her that when I am stymied for inspiration, or confused on some matter, she will find me playing Solitaire on the computer, allowing my mind to solve the problem in the background. As this is a book about God, I have to be scrupulously honest in what I write. To be honest, sometimes I am just playing a game of Solitaire on the computer.

For Mimi's long-suffering patience with my moody nature when I am writing, and her willingness as well to indulge my dilatory writing habits, I profess my enduring love. A writer with an understanding spouse has a very great treasure indeed.

Garrett Glass

INTRODUCTION

In the Beginning, we are born. We come into this world helpless and disoriented, having a brain but one-quarter the size of that of an adult. We can move our limbs about, but not our body. We can ingest food, and excrete it, but only in liquid form. We can barely see, and we can express ourselves in only one way – by crying. We have two modes of crying: loud, and louder. To the ears of others around us, our crying comes in two forms: irritating, and more irritating.

There is something about a newborn's crying that deeply disturbs adults. A newborn's crying is one of the most unwelcome sounds a passenger on an airplane will ever want to hear. Passengers are known to demand that their seats be changed in order to be far away from a crying infant. And this, all things considered, is a very good thing. We are programmed as adults to come to the aid of a crying infant, if only to stop their crying. It is how we have survived as a species for millions of years. Crying is the only tool a newborn has at its disposal, the only way it can manipulate its world, and the only way it can obtain what it wants and needs.

What an infant wants is its food, or its diaper changed, or a pain to go away. What it *needs* is rather different. It needs to be connected to other humans in a physical and emotional way, immediately as it is born. It needs to be nurtured over time, to be introduced to the wider world beyond its crib, to be given challenges that help it develop, to be taught essential skills such as speaking. What an infant needs is love, sustenance, and knowledge. And it needs these things for a very long time, through infancy, then childhood, and then as a pre-adult, until it reaches a point when it can obtain these essentials on its own.

Love. Sustenance. Knowledge. These are an infant's birthright, an entitlement, a non-negotiable demand on their part, because their survival is at stake. It would not occur to any normal adult to deny these things to an infant. As adults, but most particularly as parents or caregivers for an infant or child, we do not think twice about fulfilling our obligations to those in our charge. We lavish love on our children. We see to it that they are warm, clothed, fed, and safe. We educate them in the ways of the world. We may tire of this work; we may complain about the burden; but ultimately we do not begrudge our children what we know is owed them   It is, after all, the only time in our life we get to play God.

Not just play God; we get to *be* God. How else do we imagine our infant, or young child thinks of us? In their earliest years, we appear to them as creators of the world. We are all powerful, we know everything, we are constantly in their lives, we love them unconditionally. This is the very definition of God. At the moment of their birth, God is the first person they meet. Is it any wonder that as we grow from infancy through childhood and then to adulthood, we always have with us in the back of our mind the impression that God exists? After all, a God-like figure certainly did exist to provide us with love, sustenance and knowledge during the earliest and most critical years of our life.

This book is about a particular sort of God – the *real God* of our infancy and childhood years, the God who then proceeds to disappear from our lives as a tangible flesh and blood figure, only to be replaced by a very similar God who is intended to help us and guide us through our adult lives. The first God is not optional: all of us share some experience with God-like figures in our lives, be they our parents or caregivers who helped us through infancy and childhood. The importance of this God is that he – or she for those who prefer to believe in a feminine God - was essential to our survival.

The second God, who appears in our later childhood, is not so much an optional figure as a provisional one. Most of us have no choice in whether we believe in this God or wish to follow his rules and participate in the religion which worships him. He is a family inheritance, a gift to us as it were from our culture. As adults, however, we have a choice as to whether we wish to believe in him or not, so in that sense he is a provisional God. But let us not underestimate the importance of this second manifestation of God in our lives. While the first God was essential for our survival as an infant and a young child, the second God is considered necessary for an equally important reason. He is selected for us by our parents, our caregivers, our community, and our culture to provide us with meaning in our lives. This is the God intended to accompany us on our adult journey through life, whether we wish him to or not.

It is a commonplace thing to describe our lives as a journey. We are born and we die. We start at point A and reach point Z, and this progression is certainly a journey of events and experiences that define our life. But none of us is satisfied with describing our life this way, of defining it as nothing more than a sequence of events that can be marked by dates and occurrences. *We want to attach some meaning to these events and experiences. We want our life to have meaning.*

But what is the meaning of this word *meaning*, which we so glibly use as a vitally important descriptor of our time spent here on earth? Meaning can imply many things. It can suggest we want to be important in some way, to stand out from others around us, to have a positive impact on others, to be remembered by others after we have died. It can also describe our internal state – how we coped with life, and overcame struggles, and developed ourselves mentally and emotionally into a better person who nurtured those qualities of goodness which were of benefit to others and of satisfaction to ourselves.

Our yardstick for measuring whether our life has meaning or not consists of generalizations we can reach regarding the lives of people we admire. These can be people close to us: a friend who is notably caring and generous with her time; someone who has used managerial and organizational skills to create a business which gives meaningful employment to others, or someone who establishes a charity which bestows benefits on the needy. History is replete with people who have lived what is recognized as a meaningful life. Mahatma Gandhi, Nelson Mandela, Martin Luther King, and Aung San Suu Kyi of Burma come to mind as modern political or social figures who made great personal sacrifices to advance the interests of people who were dispossessed and given little or no role to play in the political system which shaped and governed their lives.

For billions of people, meaning in life is derived from their relationship with God. This is the second God we have been discussing, the God determined by the culture in which we are born and most often, the religious beliefs of our parents. This God is obviously different in different parts of the world, and he often comes to us through the words of men who are anointed or who sometimes appoint themselves to speak to us on behalf of God. The meaning we can obtain from a relationship with God is therefore rarely pure. It is predefined by religion, and further interpreted by religious leaders in whom we must place our trust because it is they who define God and what it is God demands of us in order to live a meaningful life.

When we have so many models from which to choose, how do we decide which one suits us best in order to give meaning to our life? If we seek religious or spiritual meaning, how do we know if it is the right path for us, especially when in most cases we had no real choice in the path we are following? Just because we grew up putting our faith in Buddha, or Allah, or Jesus Christ, does not mean that we are required to continue our life's journey with that same path. Most people, however, do just that – they would never think of abandoning what they view as their spiritual birthright, a belief system that their family has followed for many generations.

Increasingly, however, some people are indeed abandoning their childhood faith, and among their number are those who are asking some very deep questions about God. They come to a point in their life where they not only doubt that their life's journey needs to follow the footsteps laid down in advance by religion and by God; they are asking who is this God they have been taught to follow? Does he even exist? If he doesn't exist, what else is there to give meaning to the human experience?

We live in a time when they are many new and interesting ways to investigate these questions about God. There clearly are other paths any of us can follow that are not religious or spiritual, but which can imbue our life with as much significance and meaning as any of the traditional, religious paths. We have more options open to us than simply following the journey of meaning laid out for us in our childhood.

Multiple Paths to Meaning

Consider this book a short side-trip on this journey, a way for us to identify where we best can find meaning in our life. The answer may be found in a relationship we desire to have with God. But as we undertake such a relationship, we have an advantage that neither our parents nor any of our ancestors had: *we know much more about where God comes from.*

The first part of this book looks at that question, from the knowledge obtained through modern science and academic research. We investigate the discoveries about God-belief made in modern psychiatry, anthropology, neuroscience, evolutionary science, cognitive behavior, and psychology. Some of these discoveries have been made as recently as the past twenty years, and they suggest that we are just at the beginning of an important age of research into the origins of God and the reasons why belief in God persists.

Our goal in this part of the book is not to ask whether God exists. In one very specific sense, God always exists, as long as there is someone to write about him and someone like yourself who can read about him. At this very moment, your brain is summoning up thousands of memories and impressions of God, formed by a lifetime of experiences you have had reading about, discussing, or even praying to God. As you read about God in this book, you will be creating images of him in your mind, and these images may well appear real to you. Our review of recent research on God will reveal at what point in your life these images were first created, where in your mind these images reside, and why they can be important to you throughout your adult life. At the end of Part One of this book, we develop a model that describes this process of creating God in our minds.

In the second part of this book, our journey takes us beyond science, and beyond the physical world, into the world of philosophy, theology, and religion. This is the world we must investigate if we wish to address a question that is an important follow-up to our research on the God we can create in our mind. That question is: can God exist in a real, material sense, external from our mind and body? Can he be a physical being, or an invisible spirit, or a conceptual entity that serves as pure being itself, lending his essence to the creation of the universe and of all that is in it?

If our answer is negative or leads to significant doubts as to God's material and external existence, then we must ask, how was the universe created? Does it have a purpose? Moreover, how important is it for us to have answers to these questions? One of the benefits of religion is that it comes ready made with answers as to how the universe came into existence. God willed it so, is the most potent of these religious answers, and the most satisfying at a surface level, because religion does not permit us to doubt the answer. This is why we will spend some time in Part Two of this book investigating where religion comes from, how it operates, and what values it provides. Just as important, we will study the often-overlooked negatives of religion.

We will not be discussing the usual criticisms of religion, that it fosters wars and promotes superstition, for example. Other social structures that are not religious in nature do the same thing, and there are many other books which look at these problems in detail. Our purpose is different. We want to understand *why* religion exists, what powers it holds, how it connects to the God intended to give meaning to our life, and what the costs are for following a religion. In this way, we can make some judgments necessary to guide us in the discussion we pursue in the last section of this book.

Part Three investigates alternative journeys in life, and other paths we may take that are secular in nature but which still provide the meaning we seek on our journey. We begin this section with a discussion about good and evil, and about the forgotten part of that formula: misfortune. Most people who believe in God find it logical to accept that good and evil are the two primary forces of morality operating in the universe, because this is what religion often teaches. But for those who do not follow a religion nor hold a strong belief in God, dividing the moral realm into good and evil is not satisfactory. Some discussion is necessary regarding the concept of misfortune, characterizing bad events which occur to us without any purpose behind them, and which are the consequence of living in an indifferent universe which often acts in random ways.

Once we understand the three forces of good, evil, and misfortune at work, we are better prepared to analyze the religious importance of prayer and faith. Religious people seek out prayer for many reasons: to obtain relief from stress, as an example, or to find comfort in the presence of a "higher power." The benefits of this type of prayer are also available to those who are not religious, but in order for them to be non-religious, they need to confront if necessary the defenses of God-belief that constitute faith.

In this book, faith is defined as a series of mental processes which help a believer in God to ward off doubts regarding his existence. On this basis, atheism is the opposite of faith, in that it too is a process, to discard these mental defenses. Because we define atheism as a process, it is not simply disbelief in God and his existence. Atheism encompasses many roads of discovery, from doubt to agnosticism to what we call in this book atheism proper, the consequence of which is a complete denial of God's existence.

It is a common assumption that people who are not religious, especially atheists (all of whom are often assumed to deny the existence of God), can have no meaning in their life. "They believe in nothing," is what is typically said of them. An atheist is said to have no hope, because they do not look forward to an afterlife. They die, their body turns to dust, and they are forgotten. Religious people console themselves that they have a glorious future ahead of them when they die, and they feel pity for those who willingly reject the potential to spend eternity with God.

The Right Sort of Knowledge

If this were true, atheists should indeed be pitied. But it is not true, and *why* that is the case is one of the subjects we explore in Part Three of this book. The assertion made in this part of the book is that atheists of any sort, even those who are most vehement in their denial that there is a God or an afterlife, are able to meet their death with the same sense of hope and satisfaction as a religious person, *if they have lived their life well.* It might surprise you that what it takes to live your life well is nearly identical for a believer as it is for a non-believer, but that is one of the important realizations we come to at the conclusion of this book.

We said at the start of this introduction that an infant demands, and is entitled to, love, sustenance, and knowledge. Of these three necessities of life, the first two will come immediately to mind. Most of us, however, may wonder how knowledge fits into the picture. We don't appreciate that when we spend time playing with an infant or child, we are actually educating them, because every interaction with them has some educational value. Nor do we usually think of knowledge as one of the necessities of life, other than perhaps in the limited sense of knowledge gained in school and college.

That is practical knowledge, useful to help us support ourselves in this world as adults. The important knowledge that we require, however, is that which helps us obtain meaning in our life. We often tend to think that this type of knowledge comes from rabbis, ministers, imams, spiritual counselors, and other experts on the inner life of mankind. It is true that these people can be very valuable in helping us find our way not just in this world, but in this universe. They can provide us with that cosmic sense of purpose that is so elusive, but so essential if we are to approach our death with understanding and courage.

Yet too often we fall into classroom mode, acting only as the student being spoon-fed wisdom from our spiritual betters or life experts. We forget that as adults, we will never find a true meaning in our life if we do not seek it out ourselves, not just from experts, but from friends, from reading, from travel, from continuing our education in ways that suit us, and in the direction in which we wish to travel.

You've picked up this book because you are that sort of person curious about the more important questions in life. You're interested in finding answers for yourself, and not in a self-help book. This is not a self-help book. As the subtitle suggests, this is a *disquisition* into God, religion and similar matters. It is an inquiry, an essay here and there, a product of research, some suggestions – all in all, and effort to help you think about these subjects and work out your own answers.

In writing this book, I found myself on my own journey of discovery. I learned some things, and I learned which questions might never be fully answerable, for me at least. I learned that these topics of theology, cosmology, neuroscience, and so forth can be unnecessarily complex for our purposes. Therefore, as much as possible I've presented the material in this book in a way that all of us can understand.

But even more than that, I always kept in my mind one thing: *you, the reader, are on your own journey*, determining what it is you seek and what it is which might lead you to your own solutions on the "cosmic questions" regarding God, religion, non-belief, the origins of the universe and many more such topics. What is written here is intended first and foremost to help you in your personal quest, but not to push you in a particular direction.

Since this is a book about where God comes from, and not so much whether he exists "out there" in some external form, we'll begin our journey with different ways of looking at the God who resides in our mind. Later in the book I will give you my opinion about whether an externalized God exists somewhere in the cosmos. I don't think he does. I think he's an illusion in our mind, and I'll be upfront with my opinion, but it will be up to you to decide for yourself on that subject. I have found from religious people who have read this book that they do not change their minds easily on this subject, and it is not my purpose to nudge them to a different view. My purpose is to show believers and non-believers alike why there is value in each of these approaches.

There are many things we can all agree on. For one, we can all agree that our brain has something to do with how we perceive God, whether he is an external being in reality or not. Let us begin, then, in the beginning – with the first great student of the mind, Sigmund Freud. Like many of us, he didn't start out his career thinking about God, but he was drawn to the subject more and more, and ended up working out his own solution to these cosmic questions. His solutions might not be your solutions, but he had some novel and thoughtful things to say which have influenced many people over the past century. Freud opened the door to new ways of thinking about God, and he is holding it open for us now. Let's see what he has to say.

PART ONE

THE SCIENCE OF GOD-BELIEF

CHAPTER ONE

God, Religion, and Psychoanalytic Theory

Demons do not exist any more than gods do, being only the products of the psychic activity of man.

—Sigmund Freud

God is an Illusion

Sigmund Freud had two careers. The first was as the inventor of the science of psychiatry and the discipline of psychoanalysis. This work constituted the first part of his career, and words like "towering achievement" and "seminal insights" cannot do justice to what Freud created. These discoveries, born out of many thousands of hours of work with his patients, are what the public knows of Sigmund Freud.

Less well known are the fruits of his second career, when Freud turned his psychoanalytical theories toward social questions. He became a combination of behavioral analyst, historian, and sociologist, determined to apply his knowledge of the workings of the human mind to the workings of human society as a whole. As an example, he wondered whether it was possible for entire groups of people to develop collective illusions and delusions, just as an individual might, and how society might change as a result. This was daring work. Only Karl Marx had attempted to create a universal model for the unfolding of history, a model that predicted the future. Marx, however formidable his intellect and ingenious his propositions, succeeded only partially. There has always been the possibility that unrestrained capitalism might ultimately fail, but Marx insisted that it would fail and then be replaced by a dictatorship of the proletariat. That has not come to pass.

Freud's work, on the other hand, has stood the test of time, at least so far. We are not yet at a place in human history where religion has virtually died out, but the trajectory Freud laid out for such a possibility seems to be working slowly but assuredly, in a limited way. Christianity, at least in its home base in Europe, is facing a serious decline in the number of adherents to the religion, and Islam is riven by debilitating warfare between the two principal theologies of the religion – Sunni and Shi'ite Islam. In the case of Christianity, Freud has proven very perceptive. A population which enjoys the benefits of universal education, and which values the scientific method and the principles of the Enlightenment, will eventually see through the mythology which underlies religion.

This does not mean that a day will ever come when mankind completely abandons belief in God or religion, but it does suggest a transformation is underway which is eroding the theological underpinnings of what are known as the three Abrahamic religions – Judaism, Christianity, and Islam. It also suggests that a day might be coming when the general population will accept that the question of whether God exists in a material sense out in the cosmos is resolved, precisely because the more important question – Where Does God Come From? – is much closer to being answered.

It is precisely this second question which is being addressed by a rush of research, much of it scientific in nature now that medical science and neuroscience can give us an "inside look" into how the human mind works when it contemplates God or engages religion. What this modern research is accomplishing is a vindication of many of Freud's insights about God-belief as an illusion, which tells us yet again that Freud was a genius for extracting the insights he did solely from his work with his patients.

If Freud were alive today, he would be both impressed and fascinated by the discoveries being made in neuroscience, but he would think such discoveries complemented the work of the psychoanalysts, and that medical science could never rival or replace psychoanalysis. Freud no doubt agreed with the words of Alexander Pope: "The proper study of Mankind is Man." As the creator of the discipline of psychotherapy, Freud invented techniques for looking into the mind, and his understanding of the mind increased enormously as more people heard about him and came to him for help with their problems. It was by *listening* that Freud learned about how the mind worked, and it was from his countless hours listening to patients that Freud developed his psychoanalytic theories, creating along the way familiar terms such as neurosis, phobia, id, ego, and super-ego.

If there was one subject about which Freud might be disappointed when reviewing today's research, it is the inability of scientists to pinpoint where the unconscious mind resides. Psychoanalysts themselves do not entirely agree on the functions of the conscious, subconscious, and unconscious minds, other than to acknowledge that they are tiered in the brain. The conscious represents our active thought processes (which include memories) that we can summon at will; our subconscious represents hidden thought processes such as occur during dreaming, and the unconscious mind represents thought processes which cannot readily be summoned to the conscious mind but which do influence us in subtle and important ways. Since neuroscientists have not found direct evidence that there is a place in the brain specifically devoted to these mental processes, we will keep things simple in this book. We'll use the term conscious mind to refer to thought processes and memories which we can summon at will, and the unconscious mind to refer to those which cannot be brought to the conscious mind (except under duress), but which influence us nonetheless.

By encouraging students to learn from him and set up their own psychoanalytic practices, Freud guaranteed that his legacy would live on, even if he did not always agree with the methods or theories his own students developed about psychiatry. Carl Jung and Alfred

Adler are two of the most famous of Freud's students, and they each went their own way in creating interpretations as to how the mind worked. Many of Freud's students, for example, did not find his theories about psychosexual development, which were at the center of his psychological model, to be useful or even accurate. This aspect of Freud's work is mostly discredited today, especially his belief that infants and children develop sexual jealousies regarding their parents. Even Freud, later in his career, began to deemphasize this central aspect of his theories, as he spent more time exploring the role that anxieties over death played in child development and adult behavior.

It was during his second career, when he began to wonder how his theories regarding human psychology might apply on a broad, societal scale, that he wrote important and seminal works such as *Totem and Taboo*. Even more expansive was his work on religion and society, as represented in the book *The Future of an Illusion*, published in 1927. Here, Freud investigated why civilization found it necessary to worship a God, or gods. Freud was not attempting to establish whether God existed or not; he was a well-known atheist and spoke openly about his atheism despite the cost to his practice and his position in society by adopting such an unpopular view. He assumed the reader knew that they were reading a book written by an atheist. In *The Future of an Illusion*, Freud was much more interested in the question: Where does God come from? He looked at this question from both an historical and sociological perspective, with a view to applying his psychological theories as well:

> **The gods retain their threefold task: they must exorcize the terrors of nature, they must reconcile men to the cruelty of Fate, particularly as it is shown in death, and they must compensate them for the sufferings and privations which a civilized life in common has imposed on them.**

Here we see indications of something we have already mentioned – later in his career, Freud began to explore more seriously how our preoccupations around death affect our behavior. We will discuss this more fully in later chapters, but notice the important suggestion that the role of the gods is to appease man psychologically – to shelter him – from "the terrors of nature." Freud then says:

> **And thus a store of ideas is created, born of man's need to make his helplessness tolerable and built up from the material of memories of the helplessness of his own childhood and childhood of the human race.**

Freud conflates our own childhood with that of everyone else's childhood – the "childhood of the human race." In his survey of civilization, he has begun to see that man's early history constitutes a form of childhood, and that as civilization develops, mankind will progress away from reliance on the gods and religion. It is, in fact, the basic premise of *The Future of an Illusion* that the illusion, which is belief in God, will eventually die out as mankind continues to advance scientifically and through increasing application of reason and logical thinking.

Depending on where you live, this is certainly a debatable point. In the U.S., religion is still as strong as it was in Freud's time, though in Europe he would be pleased to note that his prediction seems to be coming true. But the key element of his thinking is his brief reference to man's "memories of the helplessness of his own childhood." Freud is very clear that God-belief results from that helplessness, which then leads to his "store of ideas" that lead to the necessity or inevitability of a god and religion. These religious ideas are that:

1. Life serves a higher purpose, a perfecting of man's nature.
2. Man has a spiritual part that constitutes his soul, which will detach from his body.
3. "Everything that happens in this world is an expression of the intentions of an intelligence superior to us."
4. "Death itself is not an extinction...but the beginning of a new kind of existence."
5. In the end, good will be rewarded and evil will be punished.

This short list serves as a summation of all that religion has to offer mankind. We will explore these ideas more fully in Part Two of this book. For now, let us return to the question of the development of God-belief in childhood. Freud had one more important comment to make regarding this subject:

> **We turn our attention to the psychical origin of religious ideas. These, which are given out as teachings, are not precipitates of experience or end-results of thinking: they are illusions, fulfilments of the oldest, strongest and most urgent wishes of mankind. The secret of their strength lies in the strength of those wishes. As we already know, the terrifying impression of helplessness in childhood aroused the need for protection – for protection through love – which was provided by the father; and the recognition that this helplessness lasts throughout life made it necessary to cling to the existence of a father, but this time a more powerful one.**

It is important to stress first that Freud defines these "urgent wishes of mankind" as illusions. He invented the term as it applies to psychology, and defined it as a belief in something that does not correspond fully to reality, and which has an element of "wish-fulfillment" included in the belief. The belief in God, by this definition, is not consistent with reality – there is no God in Freud's interpretation of reality. But the belief in God does have a critical aspect of wish-fulfillment, in that the believer strongly desires God to exist, and this desire arises from a childhood need for protection. In Freud's terms, belief in God is not a "delusion," because that would be a view of the world completely disoriented, and not at all connected to reality. There is no wish-fulfillment component of a delusion.

The second point to emphasize is Freud's argument that the child is protected by the father, which creates a lifelong desire for a powerful father-figure. Not all of Freud's disciples concurred with his emphasis on the relationship of the child to the father, since it left a secondary role for the mother. Eighty years later, we have come to know much more about the importance of the mother-infant bond, and its primacy in the development of the infant even when it is in utero.

We will explore this topic in a later chapter, but for the moment, let us acknowledge that Sigmund Freud set the stage for much of what is to follow in this book. He, along with the great 19th century philosopher Ludwig Feuerbach, was the first to see how God-belief operated on the human psyche, and where God-belief came from. He did not need the modern tools of medical science to reach his conclusions about God and religion. Nor do we need to take a stance on whether the "study of man" through psychoanalysis is superior to the knowledge gained about the human brain from modern science. We can do very well indeed by learning from both disciplines, but in order to understand what these discoveries are, we first have to know a bit more about how the human brain works.

In the next chapter, we will take a brief tour of the structure and functions of the human brain. Other than some Latin words which are easily explained, there is nothing fancy here or overly-scientific. There is just enough for you as a reader to have the reference points necessary to understand the neuroscientific work referenced in other chapters of Part One of this book. More than likely, you will come away from this brief survey with a sense of awe regarding this most ingenious organ which defines what it is for us to be a member of the species homo sapiens. *Let us look then, at what makes you, you.*

CHAPTER TWO

The Greatest Living Organism

In each of us there is another whom we do not know.

– Carl Jung

Basic Functions of the Brain

Any time we read about the human brain, we are doing something both unique and bizarre. We are using our brain to comprehend itself. We each have an identity – a self (or ego, as Freud called it) – which can contemplate itself. We can enter into a relation with ourselves, think about our desires, our joys, our impulses, and why we not only do the things we do, but come up with the thoughts we have. It's an endless investigation of thought contemplating where that thought came from, what action resulted, how that action generated more thoughts, and so on.

The past twenty years have witnessed an explosion in study and knowledge of the physiological mechanics of the human brain, and we are living in this exciting time. We can say, for example, that our complex thoughts come largely from one part of the brain: the *Frontal Lobe*. We know the *Frontal Lobe* is the source of our higher reasoning because neuroscientists have tools with which to study the brain in great detail. Some of these tools are medically invasive, in which a probe is used during brain surgery to stimulate certain parts of the brain, and since the patient is awake during such surgery, they can respond to questions that help the surgeon pinpoint where exactly the brain is malfunctioning. These are vitally important questions for the surgeon, who can avoid damaging any healthy parts of the brain during surgery. The questions are, amazingly, asking the brain through its owner, the patient, to answer questions about itself.

Other neuroscience tools are non-invasive[1], such as the commonly used Magnetic Resonance Imaging machine, or MRI. If you haven't experienced one of these examinations yourself, you've likely watched scenes on television of patients being wheeled into

---

1   Except for "MRI's with contrast," where a material is injected through an IV tube into the bloodstream to allow for an enhanced image.

one of these machines, which takes three-dimensional images of internal organs. Since the early 1990's, neuroscientists have been able to use the *functional* MRI, which specializes in brain and upper spinal cord imaging. The *f*MRI is an important resource for neuroscientists, because it can also show which parts of the brain have increased blood flow when a patient is given a cognitive task – such as answering a questionnaire which explores their religious beliefs. We know now, for example, which areas of our brain "light up" with increased blood flow, or usage, when we are praying to God, or meditating.

In Part One of this book, we will be reviewing how our brain creates for us an impression of God, so that we can summon him at will, talk to him, feel his presence, sense his love for us, visualize him as a human, and externalize him as a person who exists separate from us. This is the God of our mind, who seems very real to us if we wish. Even if we don't believe that God exists "out there" in the real world beyond our mind, our emotional and mental makeup has a template for a God-like figure who will always love and protect us. To understand how we create God, we need to know about some of the basic structures and areas of the human brain, so in the next few pages I will provide for you a crash course on the brain's anatomy. Don't worry! You do not need to be a neuroscientist to understand these terms, and you don't need a course in biology or medicine to comprehend how the brain works. In fact, I guarantee after reading the next few pages and looking at the charts, you will appreciate why the human brain is described as the most marvelous and complex living organism on the planet.

We've already described one important area of the brain: the *Frontal Lobe*. Let's put some definitions to these words and describe a few more that will round out our basic understanding of the brain. First, though, we should cover an important fact that is obvious if you look down from above at a human brain: there is a left part, and a right part. These are the two hemispheres of the brain, and they are in many respects mirror images of the other. As just one example, there is a small section of our brain above our left ear which perceives words and makes sense of them, without which we could not produce sentences or paragraphs. On the right side of the brain, in the corresponding spot, is a small section which perceives musical tones, which are the basic building blocks of musical phrases such as melody, and musical chords which produce harmony. This is the mirror image aspect of the brain at work: the Left Hemisphere comprehending the basic building blocks of language, and the Right Hemisphere comprehending the basic building blocks of music. The entire brain works this way in many respects, such that the Left Hemisphere is often described as the analytical part of the brain, and the right side described as the creative part of the brain. These descriptions of the two hemispheres do not always hold up; there are many areas of the brain overall that deal with analytical thought and language, even though the Left Hemisphere is usually described as the province of these two functions. Here is a view of the brain's *generalized* hemispheric functions, looking from above at the two hemispheres:

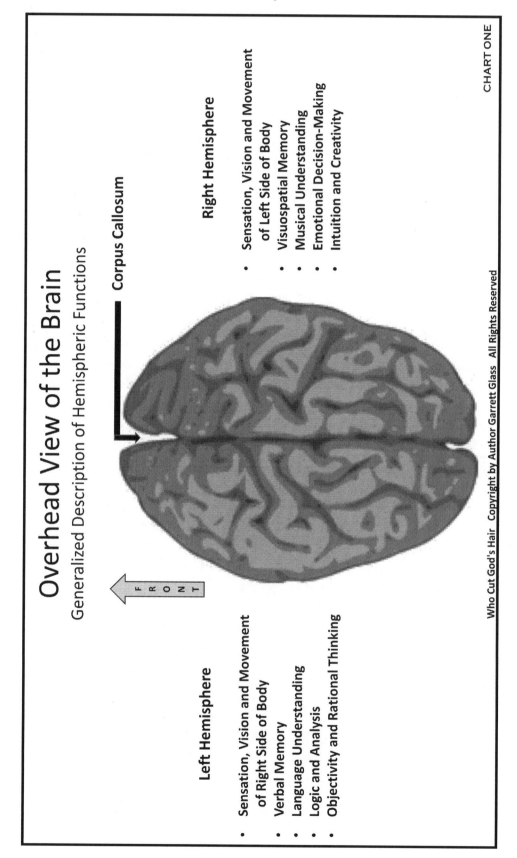

Chart One: The Two Hemispheres of the Brain

Overhead View of the Brain
Generalized Description of Hemispheric Functions

Corpus Callosum

FRONT

Right Hemisphere
- Sensation, Vision and Movement of Left Side of Body
- Visuospatial Memory
- Musical Understanding
- Emotional Decision-Making
- Intuition and Creativity

Left Hemisphere
- Sensation, Vision and Movement of Right Side of Body
- Verbal Memory
- Language Understanding
- Logic and Analysis
- Objectivity and Rational Thinking

CHART ONE

By the way, that ridge right down the middle of the brain is called the *Corpus Callosum*. It consists of bundles of fibers which connect the two hemispheres in important ways, to make sure the mind and body are operating in a coordinated matter. These fibers are very compact and tough compared to the soft gray matter that constitutes the folds on the surface of Left and Right hemispheres, which is how the *Corpus Callosum* got its name. It means "tough body" in Latin.

Now let's talk about the many bulges of "gray matter" that cover most of the brain. This is the **Cerebral Cortex,** which other animals have, but in their case the cerebral cortex is very thin. Ours is enormous and highly complex in comparison. The **Cerebral Cortex** is the part of the brain that makes us particularly human. It is where we do our complex reasoning and higher thinking, especially in the front part of the brain, called the *Frontal Lobe*. But our ability to reason about the world depends first on our ability to see the world, and there is a whole other lobe in the **Cerebral Cortex** devoted to that purpose – the *Occipital Lobe*, located on the lower back part of our brain, one half in the Left Hemisphere, and one half in the right. What about our hearing? That is regulated by the *Temporal Lobe*, one on each side of the brain, located over our ears. The two *Temporal Lobes* also process memories, and interact closely with the main storage area for memories, the *Hippocampi*. There is a left and a right *Hippocampus*, located within each *Temporal Lobe*. The *Occipital and Temporal* lobes take care of our sense of sight and hearing, but what about our sense of touch? That is governed by the *Parietal Lobe*, located on the upper back part of our brain, covering each hemisphere. The *Parietal Lobe* gives us our ability to understand not just touch, but the space around us. It allows us to physically navigate the world. The *Parietal Lobe* also allows us to sense and interpret temperature, taste, and pain.

The **Brain Stem** is found at the top of our spinal cord and is situated in the middle of the brain, where it directs signals from our body to our brain, and back to our body. The **Brain Stem**, for example, regulates our heart beat, pulse, blood pressure, and respiration. It also has the capacity to store memories.

You may have noticed above that some words are in bold letters, and some words are in italics. There is a purpose to this. The human brain has four elementary components, the result of millions of years of evolutionary development. These four basic components are in bold letters, and you have now learned about two of them: the **Cerebral Cortex** and the **Brain Stem**. Each of these components have sub-parts, and these words will be in italics. I will use this format only in this chapter, so that you can get used to thinking of the brain on three levels: the hemispheres, the four elementary components, and the subparts to these components.

The **Cerebral Cortex** itself has four critical sub-parts, called lobes. Here is a picture of the *Frontal, Parietal, Occipital,* and *Temporal* lobes, all part of the **Cerebral Cortex. The Brain Stem** and **Cerebellum** are shown as well:

Chart Two: Left Hemisphere View of the Cerebral Cortex with Cerebellum and Brain Stem Also Shown

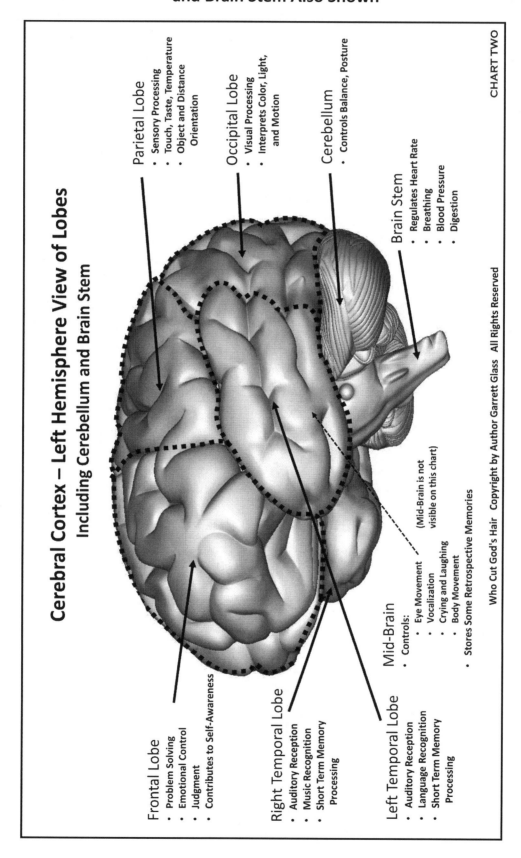

Cerebral Cortex – Left Hemisphere View of Lobes
Including Cerebellum and Brain Stem

Parietal Lobe
- Sensory Processing
- Touch, Taste, Temperature
- Object and Distance Orientation

Occipital Lobe
- Visual Processing
- Interprets Color, Light, and Motion

Cerebellum
- Controls Balance, Posture

Brain Stem
- Regulates Heart Rate
- Breathing
- Blood Pressure
- Digestion

Frontal Lobe
- Problem Solving
- Emotional Control
- Judgment
- Contributes to Self-Awareness

Right Temporal Lobe
- Auditory Reception
- Music Recognition
- Short Term Memory Processing

Left Temporal Lobe
- Auditory Reception
- Language Recognition
- Short Term Memory Processing

Mid-Brain
- Controls:
  - Eye Movement
  - Vocalization
  - Crying and Laughing
  - Body Movement
- Stores Some Retrospective Memories

(Mid-Brain is not visible on this chart)

CHART TWO

We should note that there are other lobes in the brain as well, and one of them, the *Insular Lobe*, is in the interior of the brain and is not shown on the chart above. The *Insular* lobe is connected to many parts of the brain, and is believed to play some role in consciousness and self-awareness. Among mammals, only the Great Apes and humans have an *Insular Lobe*, allowing both species to generate feelings of empathy, as well as other complex emotions. We will discuss these important emotional processes when we come to the **Limbic System**.

The **Brain Stem** is the structure leading downwards from the bottom of the brain in the picture above. It is the part of the brain that connects to the spinal cord. As we mentioned above, the **Brain Stem** is the basic mechanism by which the brain sends and receives signals with the rest of our body, so that we can walk, digest food, pump blood through our heart, use our hands and fingers, enjoy sex, give birth (only half of us are privileged to do this), and feel pain. The brain stem is usually described as the most primitive part of the brain, because it is the part of the brain that we have in common with almost all other vertebrates, or animals with a spinal system like ours. Lizards and other reptiles have brain stems, and therefore this is the part of our brain that is sometimes referred to as our "lizard brain," which means it contributes to our most primitive impulses that were useful before our species had language or complex thought. This includes our instincts to sense danger, our willingness to fight (even those of our own species), and our basic, less sophisticated interpretations of what we are seeing and hearing.

Most of this functioning is located at the very top of the **Brain Stem**, and this area is called the *Mid-Brain,* because it is positioned in the middle of our brain, so that it can communicate with the rest of the brain. As such, the *Mid-Brain* controls eye movement, our ability to vocalize, and our capacity to laugh and cry. It allows us to use our arms, hands and legs in a purposeful way, and damage to the *Mid-Brain* is linked to problems such as Parkinson's Disease. The *Mid-Brain* is the part of the **Brain Stem** that has the capacity to store long term, retrospective memories. These may include memories from our infancy that are inactive and no longer processed by the conscious mind.

We now have what we need to know about the **Cerebral Cortex** and its four lobes: the *Frontal* (complex thought and reasoning), *Parietal* (touch and spatial coordination), *Temporal* (sound), and *Occipital* (sight). We also know about the **Brain Stem**, the link between the brain and the body, and also the source of our sense of taste. There are four fundamental parts of our brain, and we now know about two of them: the **Cerebral Cortex**, and our primitive brain, known as the **Brain Stem**. The third fundamental part of the brain is shown on the chart above and is a mini-brain of sorts. It looks like a miniature version of the overall brain, so it is called the **Cerebellum** (Latin for small brain). This is also a rather primitive part of our brain that was developed before the **Cerebral Cortex** gave us higher reasoning powers. The **Cerebellum** is located at the lower rear of our brain, connecting to the **Brain Stem**, and below the *Occipital Lobe*, and its important role is to give us our upright posture and sense of balance. This is why it is considered a primitive part of the brain: when our earliest ancestors millions of years ago distinguished themselves from other members of the primate family, it was not just from having language or higher reasoning, it was from the ability to walk upright, run, and leap. These powers are controlled in the **Cerebellum**.

Now that we know about the **Cerebellum**, we can add that to the list of the four fundamental components of our brain. That gives us the **Cerebral Cortex**, the **Brain Stem**, and the **Cerebellum**. What is the missing fourth component? That is called the **Limbic System**, and that is the last stop on our journey to survey the fundamental structure of what is the most complex living organism ever created on earth. The **Cerebral Cortex** gives us thought and processes our sensory sensations (sight, sounds, etc.). The **Brain Stem** links the brain to our body and controls basic functions like sleep, alertness to danger, body temperature, pulse, and heart rate. The **Cerebellum** gives us our unique posture and balance. The **Limbic System** gives us that one other vital aspect of humanness, and that is our emotional makeup.

Through the **Limbic System**, we process happiness, pleasure, love, empathy, altruism, and negative emotions as well, such as fear, disgust, anger, and rage. Another critical feature of the **Limbic System** is that it processes and stores our memories, which are of vital importance to our thinking abilities. The **Limbic System** has a number of sub-parts which play a critical role in our emotional perception of God, and we will refer to these important sub-parts by their Latin names (the language of the first scientists to study the internal workings of the brain). These are the only terms you will need to know for our purposes in discussing God and the human brain: the *Cingulate Gyrus*, the *Nucleus Accumbens*, the *Amygdala*, the *Striatum*, the *Thalamus*, and the *Hippocampus*. Here is a chart describing these parts of the **Limbic System**, and how to pronounce these words. I've included as well the functions of the major lobes of the **Cerebral Cortex**:

Chart Three: Main Functions of the Cerebral Cortex and the Limbic System

Main Functions of the Cerebral Cortex and Limbic System

Brain Section	Pronunciation	Function
Frontal Lobe	FRONT-al LOWb	Higher Reasoning
Parietal Lobe	Pah-RYE-ah-tal	Provides Spatial Orientation, Touch
Occipital Lobe	Oh-SIP-pi-tal	Processes Vision
Temporal Lobe	TEMP-or-al	Processes Hearing
Cingulate Gyrus	SING-yu-let JAI-rus	Intermediates between Cerebral Cortex and Limbic System, Promotes Learning and Motivation
LIMBIC SYSTEM		
Thalamus	THAL-ah-mus	Source of Positive Emotions, Communicates with Mid-Brain, Regulates Consciousness and Sleep
Hippocampus	HIP-po-cam-pus	Processes Memories
Amygdala	Ah-MIG-dah-lah	Generates Fear, Anger, Flight Impulse
Striatum	Strigh-EIGHT-um	Moderates Emotional Responses
Nucleus Accumbens	NEW-klee-us Ak-KUM-bens	Issues Chemical Rewards such as Dopamine

In this next chart, we look at a cross-section view of the brain, which reveals the **Limbic System**, the **Brain Stem**, and the **Cerebellum**, which are the three major components of the brain along with the **Cerebral Cortex** and its four lobes. In this chart, only a very small portion of the *Temporal Lobe* is shown.

Chart Four: Cross-Section View of Left Hemisphere of the Brain, with Brain Stem and Cerebellum

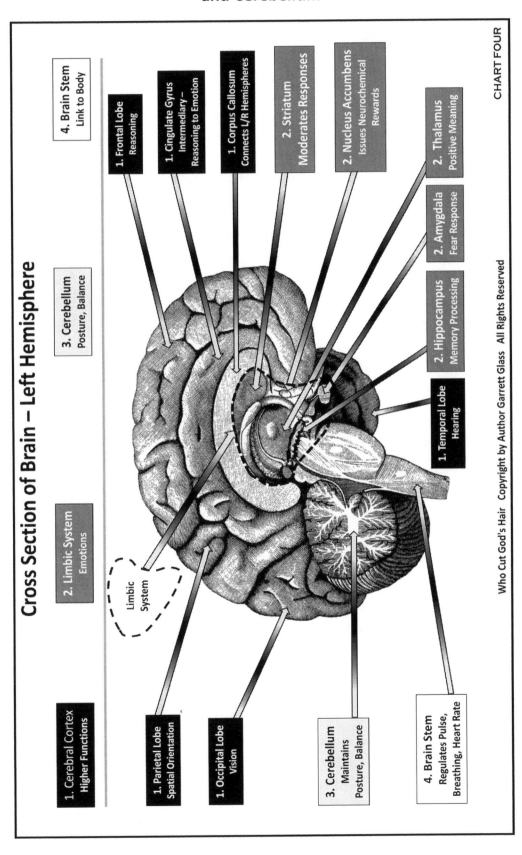

Cross Section of Brain – Left Hemisphere

4. Brain Stem
Link to Body

3. Cerebellum
Posture, Balance

2. Limbic System
Emotions

1. Cerebral Cortex
Higher Functions

1. Frontal Lobe
Reasoning

1. Cingulate Gyrus
Intermediary –
Reasoning to Emotion

1. Corpus Callosum
Connects L/R Hemispheres

2. Striatum
Moderates Responses

2. Nucleus Accumbens
Issues Neurochemical
Rewards

2. Thalamus
Positive Meaning

2. Amygdala
Fear Response

2. Hippocampus
Memory Processing

1. Temporal Lobe
Hearing

1. Parietal Lobe
Spatial Orientation

1. Occipital Lobe
Vision

3. Cerebellum
Maintains
Posture, Balance

4. Brain Stem
Regulates Pulse,
Breathing, Heart Rate

Limbic System

CHART FOUR

Who Cut God's Hair Copyright by Author Garrett Glass All Rights Reserved

— 39 —

With the human brain being as complex as it is, nothing is ever cut and dried in these descriptions. There is considerable duplication of function in the brain, and not just in the hemispheric divisions. For example, the *Temporal Lobe* maintains memories of sounds that we hear and that are important to us, and these are shared with the **Limbic System**, and adjusted constantly. At all times, the brain is communicating from one part to another, and across one hemisphere to another. There is such complexity to the brain that we can consider neuroscience to be a discipline that is still in its youth, with much more to learn.

Having said that the brain constantly communicates from one hemisphere to another, there are exceptional cases where people are born without, or have suffered damage to, the *Corpus Callosum*. Sometimes brain surgeons deliberately sever the *Corpus Callosum*, for instance with patients who suffer from daily epileptic seizures. The surgery reduces the number of seizures dramatically, but effectively such individuals have two separate brains that cannot communicate with each other. The individual can lead a normal life, but there are odd consequences to such surgery. Noted neuroscientist V.S. Ramachandran of the University of California at San Diego, conducted experiments with an individual with a severed *Corpus Callosum* to see how each hemisphere functioned independently. He asked the subject, "Do you believe in God?" One hemisphere of the brain answered the question with an emphatic "Yes." The other responded "No."

This is compelling evidence that God-belief is created by and located in the human brain. As is often said in Evangelical Christianity, "God does not make mistakes." Religion assures us that God wants all people to come to him with faith in his existence and power. He would not possibly make someone who entertains belief and non-belief in God at the same time, with equal fervor, and since such people surely exist, their existence argues that God does not exist, at least not in a material, external sense.

The marvelous thing about these discoveries of the human brain is that they are occurring now, with more research results arriving daily. Some neuroscientists have worked specifically on questions regarding God-belief and religious convictions. One such neuroscientist is Dr. Andrew Newberg, who along with his research partner Mark Waldman, conducted an extensive investigation on these important questions. In the next chapter, we explore the conclusions Dr. Newberg and his colleague reached. Now that we have our preparatory work done on the basic functions of the brain, these research results will be both understandable and surprising. God, it turns out, operates in many different areas of our brain. Come read about how our brain creates and perceives God, and how it is we develop the belief that God is real.

CHAPTER THREE

How the Human Brain Perceives God

The only difference between me and most people is that I'm perfectly aware that all my important decisions are made for me by my subconscious. My frontal lobes are just kidding themselves that they decide anything at all.

— Rex Stout

Recent Discoveries in Neuroscience

We'll proceed next with a model created by Andrew Newberg, a neuroscientist who with his colleague Mark Waldman did research on the brain and its relationship to God and religion. Their book discussing this research is titled *How God Changes Your Brain*.

Andrew Newberg is a licensed physician who also practices as a neuroscientist. As noted earlier, one of the chief tools of a neuroscientist is the functional Magnetic Resonance Imaging machine, or *f*MRI. This is a non-invasive device that takes three-dimensional images of the brain, and has revolutionized the study of the brain in the past 20 years. In scientific terms, that is a very short period of time, and we are only at the beginning of a fascinating period of discovery. Already, through the use of the *f*MRI, scientists like Dr. Newberg have established breakthroughs in our understanding of what happens in the mind when God and religion are thought about, read about, or discussed.

Dr. Newberg also used positron emission tomography, known as a PET scan. This device requires that the subject have injected into their bloodstream a radioactive dye which can be used by the PET scan to monitor among other things, blood flow and oxygen consumption in the brain and elsewhere. The PET scan is better suited for certain tests. For example, it is possible to analyze someone during the practice of meditation with a PET scan, and even to see which lobes or interior parts of the brain are activated (or deactivated) by the practice of meditation. As these observations are combined with what has been learned separately about the individual components of the brain, a map can be created which shows what is happening in the mind when God, religion, or spirituality are contemplated.

Dr. Newberg discovered, for example, that when someone is in deep meditation, the Parietal Lobe virtually shuts down, and brain function shifts to the Frontal Lobe, as the meditator focuses intently on the object of their devotion. Normally, the Parietal Lobe helps us differentiate our body from the rest of the world, but in intense meditation, the boundaries between our body and the outer world are reduced or eliminated. The subject feels intimately connected to the outer world – to a greater good, or to what some describe as God himself. Scientists call this phenomenon "ego dissolution," – the sense of self, or ego, dissolves, allowing a person to feel intimately connected in space and time to the outer world. Many who have experienced ego dissolution describe it as a sense of Oneness with the universe, and they say it has changed their life forever. Some of these individuals say that the experience of ego dissolution has made them more religious and more inclined to believe in God, while at the same time they are less anxious about life.

What we are discovering through science is that *we humans can create ego dissolution*, which means we can create the conditions under which a strong belief in God has the potential to arise. In one such study, neuroscientists administered controlled doses of psilocybin to test subjects. Psilocybin is popularly known as "magic mushrooms," and many users of psilocybin have reported a sense of Oneness with the universe, which is the mental condition known as ego dissolution. The neuroscientists were able to monitor the brain activity of users of psilocybin, and they found that activity in the Parietal Lobe was significantly depressed by the drug. This is similar to the result Dr. Newberg observed in nuns when deep in prayer, or in those who meditate for the purpose of achieving Oneness with the universe – they also show reduced Parietal Lobe functioning.

A similar study was recently published in 2016 by the Imperial College London, describing the results from monitoring test subjects who were given controlled doses of the hallucinogenic drug LSD. Using an ƒMRI, which looks at blood flow in the brain, and other devices which measure electrical activity in the brain, researchers found that communication with the separate components of the brain is significantly enhanced under LSD. In a sense, the whole brain experiences ego dissolution. The effect is also similar to the condition known as psychosis, in which normally unconnected parts of the brain are capable of communicating with each other, allowing for a more child-like state of being. The Occipital Lobe in LSD users, for example, communicates actively with the Hippocampus, which stores memories. This allows users to tap into their memories and experience them in a visual sense, which is one aspect of the hallucinatory effect of LSD.

Not everyone's experience with psilocybin or LSD is positive, and for that reason the neuroscientists who conduct these experiments use carefully controlled doses of these substances under tightly-monitored conditions. The results are promising enough that neuroscientists and psycho-pharmacologists, who specialize in the development of drugs for psychiatric use, are exploring both psilocybin and LSD as potential drugs for the control of depression, addictive behaviors, and for what is referred to as "end-of-life anxiety." Terminally ill patients in hospice care have been given psilocybin, and many of these patients have reported much less fear of death, and more of a sense of Oneness with the universe now that their life is coming to an end. Those who are religious and in hospice care report that psilocybin brings them closer to their God and more eager to explore what they believe to be the afterlife. Science, in other words, has learned through psychoactive drugs how to help humans summon up God.

Understanding how the phenomenon of ego dissolution works was but one aspect of Dr. Newberg's research with individuals who were asked to think about or visualize or read about God. In studying hundreds of individuals, Dr. Newberg came up with many other observations of how the brain perceives God. Here is a summary of his research describing how different parts of the brain help create and define God:

Frontal Lobe – *Creates and integrates all of our ideas about God – positive or negative – including the logic we use to evaluate our religious and spiritual beliefs.*

Occipital Lobe - *Identifies God as an object that exists in the world; allows us to visualize him.*

Parietal Lobe - *Establishes a relationship between the two objects known as "you and God." It places God in space and allows us to experience God's presence.*

Temporal Lobe – *Creates a voice for God. Allows us to enter into a personal conversation with God.*

Hippocampus – *Stores memories of God, reinforces memories over time.*

Cingulate Gyrus – *Creates a compassionate God.*

Thalamus – *Gives emotional meaning and a sense of reality to our concepts of God, especially a loving God. Promotes altruism.*

Amygdala – *When overly stimulated, creates the emotional impression of a frightening, authoritative, and punitive God.*

Striatum – *Inhibits activity in the Amygdala, allowing us to feel safe in the presence of God.*

Chart Five maps out the areas of the brain which create and perceive God, according to the Newberg/Waldman Model:

Chart Five: Dr. Andrew Newberg Model for the Brain's Perception of God

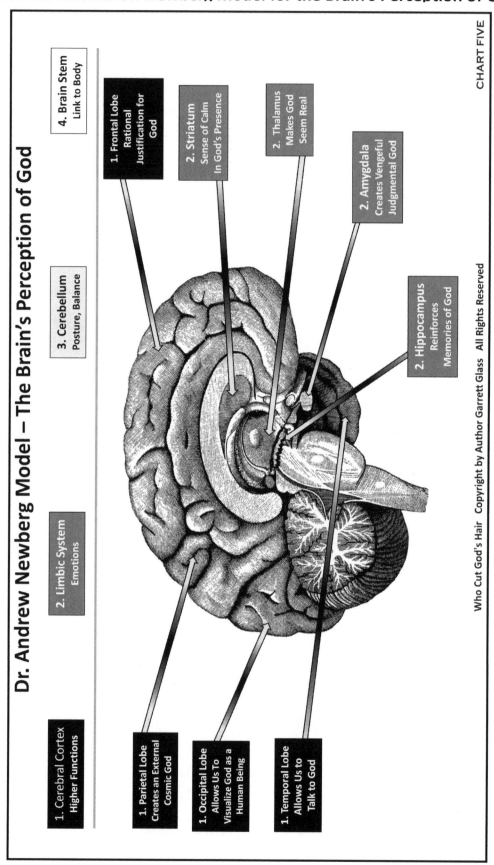

Dr. Andrew Newberg Model – The Brain's Perception of God

1. Cerebral Cortex
Higher Functions

2. Limbic System
Emotions

3. Cerebellum
Posture, Balance

4. Brain Stem
Link to Body

1. Frontal Lobe
Rational
Justification for
God

2. Striatum
Sense of Calm
In God's Presence

2. Thalamus
Makes God
Seem Real

2. Amygdala
Creates Vengeful
Judgmental God

2. Hippocampus
Reinforces
Memories of God

1. Parietal Lobe
Creates an External
Cosmic God

1. Occipital Lobe
Allows Us To
Visualize God as a
Human Being

1. Temporal Lobe
Allows Us to
Talk to God

CHART FIVE

Who Cut God's Hair   Copyright by Author Garrett Glass All Rights Reserved

Beyond conducting *f*MRI tests on subjects, Dr. Newberg and his colleague conducted over one thousand interviews, asking subjects, including children, about God, religion, and spirituality. One of their initial findings was that a child can only grasp a concept such as God in a limited way, as a noun without complex attributes. Newberg writes:

> **But what about the word God? We believe that this concept has a specifically unique effect upon the brain, because the evidence shows that children begin with a simple concrete image of God that slowly becomes more abstract and emotionally arousing in either a positive or negative way.**
>
> - **Below the age of six, children usually draw God as a face.**
> - **From age six to ten, children draw God in the company of other people. A common theme is that of God as a king, living in a palace.**
> - **As children reach their teenage years, they are able to comprehend more abstract ideas. They tend to draw God using radiating spirals and lights, as many adults do.**

Importantly, from an early age, children invest God with human characteristics, so that they can relate personally to him. Second, they early on conceive of God as a corporeal being, with a body and mind separate from their own, just as they also understand around ages one to two that their parents and caregivers are individuals in their own right, and that they themselves as a child have their own body and their own identity. Later, when the child is able to understand abstract thoughts, they are often introduced to spiritual dimensions of religion and the spiritual and holy nature of God. Dr. Newberg notes:

> **Thus, when a child is introduced to a spiritual concept, the brain automatically gives it a sense of realness and personal meaning...God is a noun, and nouns stimulate the "where" and "what" part of the brain, specifically regions in the parietal lobe. This region is responsible for identifying and integrating shapes it perceives in the world...Thus, the brain attempts to place God somewhere within the physically observed universe.**
>
> **-Newberg and Waldman, *How God Changes Your Brain***

This is an important finding: as children, we learn early on about God, and we have a ready-made mechanism, through the brain's Parietal Lobe, to conceive of God as living outside of ourselves, as a real being with human characteristics, and located somewhere "out there." This could be out in the cosmos, since children are taught that God is the creator of the world, and he would therefore reside in a place called heaven.

Dr. Newberg's work has established many areas within the brain that help us perceive God's qualities. Much of his work focuses on the emotional qualities we associate with God. What about God's traditional, fundamental qualities? Does not the standard definition of God tell us that he is our all-powerful, all-knowing, always present, all-loving, Creator? To find out where this basic image of God comes from, we will turn in the next chapter to the evolutionary imperative for God-belief. We will go back to our moment of birth, when we first meet God directly.

CHAPTER FOUR

The Evolutionary Case for Belief in God

What is Man? Man is a noisome bacillus whom Our Heavenly Father created because he was disappointed in the monkey.

- Mark Twain

The Importance of the Mother-Infant Bond

Current research on the mother-infant bond, and how it might create a life-long desire for the presence of God, has been amply summarized in a book published in early 2016, *The Illusion of God's Presence: The Biological Origins of Spiritual Longing*, by John Wathey. John Wathey makes an interesting and compelling argument regarding the human need for a God, and he brings to this argument his skill and experience as a neuroscientist. He opens with an observation often overlooked regarding God, and then a statement regarding the basic premise of his book:

> **Few believers are interested in worshipping a god who merely set the universe in motion and then sat back to watch it like a TV show, never getting actively involved in subsequent events. Instead, the god most of them worship is a personal god, one who follows and cares about the events of their individual lives, who knows and loves them, who has the power to perform miracles, and who hears and answers their prayers. He is worthy of the hymns they sing to him, because he is omnipotent, omniscient, and benevolent.**

The author goes on to add:

> **For now I merely emphasize that religious belief appears to be a completely natural, neurobiological phenomenon.**
>
> **-John Wathey, *The Illusion of God's Presence*, Chapter 1, pg. 25**

Wathey observes that people hold a belief in God on the basis of emotion and not reason. God-belief does not withstand any test of rationality or logic, as evidenced by the fact that believers receive considerable and consistent evidence that at least half of their prayers on average go unanswered by God. Yet their belief remains unshaken, and various non-rational justifications are put forth to circumvent whatever doubts may exist about

God's inability to answer every prayer (he knows what is best for us, we did not pray hard enough, he is testing us, etc.).

One argument Wathey explores to explain this paradox is derived from the work of cognitive theorists who argue that early humans developed an ability to detect other predators, including human beings, whose intentions could have been evil. It was therefore necessary to detect other humans before they were visible, which meant detecting other humans through smell or sound, or even detecting them as a subtle intrusion on one's personal space. This capacity to sense humans before they were physically observable led to an additional important capability: to attribute human qualities to non-human objects – to see other humans all around us. We modern *homo sapiens* have inherited this capacity. Everyone today has the ability to see a human face in nature, or in material objects.

In 1976, NASA's Viking probe returned photos of the planet Mars, and one photo in particular captured the public's interest:

Chart Six: NASA Image of Face on Mars

NASA Image of Face on Mars

The "Man on Mars" image led to feverish speculation that aliens had planted this image on Mars. NASA spent twenty years trying to explain that the image was just a trick of light and shadows, and eventually additional photos of the area revealed exactly that – the Man on Mars was a rock formation that disappeared when viewed under brighter light and different angles.

Everyone could "see" the image of a human face from the photo. That much was not in dispute, and though evolutionary experts and psychologists pointed out that humans have an innate propensity to see human faces in nature, the public could not resist speculating on what the image implied. We are fascinated by the human face because it is an essential aspect of our survival as an infant to be able to read human faces. They are our link to food, clothing, shelter, and love – in short, to God-like beings who provide us all we need and desire. This is also why humans are capable of seeing the face of Jesus Christ in a piece of burnt toast, or in the bark of a tree. We deliberately seek out confirmation that God exists and will prove his existence to us even by revealing himself in a degrading way, on a piece of toast, or in an oil slick. These are examples of the tendency of humans to anthropomorphize and externalize God – to imagine God as a human figure, through our Occipital Lobe, which creates the image, and our Parietal Lobe, which sets the image physically apart from us, "out there" in the external world. If it happens that we believe God is invisible, we are still able to invest in God different human qualities. Allah, for example, has 99 human qualities ascribed to him, even though it is a profanity within Islam to visually depict Allah or his prophet.

Wathey cites the important work done by anthropologist Pascal Boyer, in his book *Religion Explained: The Evolutionary Origins of Religious Thought*. Boyer is a cognitive anthropologist, interested in how the human brain and the evolutionary process of its development contributes to human cultures, including aspects such as religion. Boyer argues that evolution has created a human instinct for a supernatural force which created and now governs the universe. This force is granted human characteristics, and appears either as multiple gods or as a single, universal god. From this god, follows religion, which in Boyer's view sprung up almost naturally within ancient tribal and agrarian societies. This instinct explains the persistent reliance on religion and God-belief by cultures around the world to explain the universe, and therefore our human tendencies to depend on classical ontological explanations for the universe (i.e., there must be a Prime Mover as creator). In Boyer's framework, because cultures around the world vary greatly, the nature of God and religion varies accordingly.

It is not that Wathey argues with or wishes to overthrow Boyer's analysis – it is that he wishes to augment it in a critical way:

- **A human infant is born helpless and, if separated from its parents, is at great risk of death from starvation, dehydration, extremes of temperature, or, in the natural state, predation.**

- **There is innate neural circuitry, both in the parents—especially the mother— and in the infant, the purpose of which is to establish and maintain a powerful emotional bond between parent and infant. This bond normally prevents separation and insures adequate parental care of the infant...**

- **In the infant, this circuitry does more than merely cause the infant to cry when cold, hungry, or alone. It constitutes an internal neural model of the infant's world. The single most important feature of this model is an innate mental image of the mother as a loving agent who wants to satisfy the infant's needs and who will do so if only the infant cries loudly and long enough. Included in this neural model is a feeling of certainty that this agent exists.**

- **The circuitry that implements the innate model of the mother persists into adulthood, but normally lies dormant.**

- **The existence of this circuitry in the adult gives rise to religious experience, and hence religious belief, especially under conditions of great physical or emotional stress that evoke feelings of desperate need and helplessness.**

> **-John Wathey, The Illusion of God's Presence, Chapter 4, pp. 61-62**

Wathey then identifies those aspects of a mother's care that matter most to an infant:

Those aspects of the mother most relevant to the infant's survival are that she:

1. **exists;**
2. **knows how to meet the infant's needs;**
3. **is able to meet those needs;**
4. **wants to meet those needs; and**
5. **will meet those needs if summoned by the infant's cries.**

> **-John Wathey, The Illusion of God's Presence, Chapter 4, pg. 62**

To buttress this argument, John Wathey presents many cases in the animal kingdom where maternal imprinting is essential for the survival of these species. We need not go through these examples here, other than to say that the mother-infant bond has been well-established among a considerable variety of species, and that medical experiments have demonstrated where this instinct is located in the brain, and how it works. Since man is as much a biological creature as any other of the higher species, we cannot be exempt from the same neural imprinting instincts, according to Wathey.

It is difficult to argue against this statement. Ask any mother, for example, what sort of instantaneous change came over her emotionally when her newborn was first placed in her arms and held to her chest. An immediate emotional bond to the infant sprang up, seemingly out of nowhere, and the infant for its part displayed an instinct to find the nipple, knowing from its own neural model that this was its source of food.

Wathey also refers to the work of John Bowlby, the developer of Attachment Theory. Bowlby studied under Melanie Klein, a student of Freud's about whom we will talk at length in the next chapter. Klein created her own school of psychology, in which she modified Freud's theory that a child's sexual obsessions with its parents will influence its behavior later in life. Her modified view stated that infantile development is related to the infant's fascination, but not obsession, with the bodies and sexual capacities of the mother and father. Bowlby wasn't interested in either of these theories. His view was that an innate emotional bond with the mother and the infant is evolutionarily beneficial, and moreover, can be observed and studied by working with mothers and their infants. From these observations he could see that an infant will play with a stranger if the mother is present, but when she leaves, most infants will start to cry in distress, only to stop when the mother returns. This behavior also shows that infants are willing to explore their world on their own, or with strangers, as long as they are secure in the knowledge that the mother is present, even if she is at a distance. This is known as secure attachment.

Bowlby's associates did follow-up research on the implications of belief in God for Attachment Theory. Psychologist Lee Kirkpatrick, in his path-breaking book *Attachment, Evolution, and the Psychology of Religion*, found a plausible explanation for God-belief using Attachment Theory. An infant who develops a secure attachment to the mother (or other caregiver) is likely to have an adult belief in a loving, omnipresent, and caring God. Such adults operate with the comfort that God is always nearby, making the world a safe place, just as when they were infants, when they were able to function well with the secure knowledge that their mother was nearby. These adults also are apt to be highly religious. Some infants do not cry when the mother leaves, or will not easily be calmed when the mother returns because they are angry at her. Observations have shown that these infants, when they reach adulthood, can be more prone to agnosticism or atheism. Indeed, in these adults, the pain they experience when leaving religion is similar to that of infants who are permanently separated from their primary caregiver.

Attachment Theory, combined with the work of anthropologists and cognitive theorists, according to Wathey, are disciplines which explain the extreme power of the instinctive desire for God, and the persistence of religion. Wathey uses the term "hard-wired" to describe the neural model between an infant and its mother, but in his view *this does not mean* that mankind is hard-wired to believe in God. He writes, "The essence of my hypothesis is that an adult's sensation of God's presence arises from the innate neural circuitry that, in infancy, initiates the infant's part of the mother-infant bond." This suggests that all of us have an innate capacity to bond with our mother, and it is this mother-infant bond which later in life can lead to an adult having a "sensation of God's presence."

In my own view, and I'll pursue this argument later in the book, it is not so much that we are forced to believe in God, which is what a hard-wired neural model implies. It is that **God-belief is the default position for humans**, many of whom naturally fall into God-belief, and some of whom successfully resist God-belief. The number of those who do not believe in God is growing over time as a percentage of the population, at least among industrialized nations, so even if we are hard-wired at birth to believe in a powerful, loving and beneficent presence, this would not force everyone to believe in God. We must therefore conclude that the neural model for an all-powerful, all-loving, all-knowing, and always-present Creator figure, does not *force* us to believe in God, but it does *incline* us to believe in God. That is why God-belief is the default position for mankind. I refer to this neural model as *Template God*, a model which all of us share as it is necessary for our survival, and which we will discuss in a later chapter.

Once the template for God is created in an infant, it eventually is processed by that area of the brain which excels at providing rational explanations for phenomena we experience in the outer world – the Frontal Lobe. In the brain of an infant or child, this element of the Cerebral Cortex is just beginning a lengthy journey of expansion and complex development, but even in the first few months of the life of an infant, the Frontal Lobe is capable of providing the infant the ability to make a basic decision, such as "if I cry, my mother will feed me." Wathey points out that over hundreds of thousands of years, humans have learned which decisions are favorable and which are not, and from an evolutionary perspective, our species clearly thrives if a strong mother-infant bond can be created at birth. The hard-wiring comes into play if an infant can immediately sense its mother's presence through smell, touch, sound, hearing, and initially, limited sight. The infant learns as well how to distinguish its mother from others and how to call for food and other help, since initially the infant is absolutely helpless. All of these cues work to the benefit of the infant and aid in its survival; hence the hard-wired nature of this decision-making.

This reinforcement process is profoundly important for the infant because survival is at stake, and it should not surprise us that the neural instinct of "ask and you shall receive" turns into a rational conclusion that "there is a figure in the universe who will give us all that we need." Therefore, it is likely we have stored in our brain, possibly in the Mid-Brain, our earliest memories of a God-like figure, which also provides us with a rational justification for the existence of such a person.

But more is needed for this process to work than just rational appreciation that crying will summon the mother, even if this understanding has become an instinct created over countless repetitions by our ancestors. A mother's touch, her voice, the sensations of sweetness and fat that are stimulated by her milk – these have to trigger a consistent neurological response if the infant is to want to return time and again to these stimulations. The neurological response, from Wathey's review of research on the matter, is a neurochemical that originates in the Limbic System, and is to be found in the Nucleus Accumbens. This is a tightly-bound, compact collection of neurons that has as its principal purpose the issuance of dopamine (and other chemicals) throughout the brain as a reward for certain behaviors. The brain has to regulate carefully the amount of dopamine that will be issued for any particular purpose. If an infant has too little dopamine generated in its body, it can suffer from mental deficiencies. Too much dopamine, as would happen with a dose of cocaine, leads to a temporary sensation of euphoria, sexual omnipotence, and energy, but the reward quickly becomes something the body intensely craves. An addict is born, by over-stimulating the Nucleus Accumbens.

In Wathey's neural model, the Frontal Lobe generates an action born of instinct, such as an infant smiling at the sight of its mother; the Nucleus Accumbens rewards this behavior with a small burst of the neurotransmitter dopamine. The mother, in turn, rewards her infant with a smile as well, and the mother receives pleasure from the interaction too. Mother-Infant bonding is reinforced through instinctual behavior and a neurotransmitter (dopamine) reward.

How do researchers know that mother-infant bonding is a function of brain activity located in the Frontal Lobe and the Nucleus Accumbens? The fMRI technology is an important diagnostic device to observe infant and maternal response to different stimuli, such as the absence or presence of the mother. Since it is not always easy or ethical to use infants as subjects of fMRI research, this research can be done through proxies, on our close genetic relatives such as monkeys, but also on laboratory rats. If the Nucleus Accumbens is deactivated in a mother laboratory rat, she will ignore the cries for assistance from her offspring, and will also lose interest in basic functions such as eating. The reward system that is managed by the Nucleus Accumbens is a powerful motivational force for fundamental instincts and behaviors in almost all vertebrates, and perhaps the most powerful of these instincts as seen across the animal kingdom is the maternal-infant bond. Nature seems to have devised a universal way for many animals to benefit from a strong maternal-infant bond for whatever time it takes for the infant to mature.

The major developmental difference between humans and any other animal is that we humans have an incredibly long development period from infancy to adulthood. Our offspring cannot exist on their own until they are, in many cases, 22 years old, if we take into account the need for higher education in order for an individual to survive and support a family in any industrialized society. Even the period of infancy, constituting the first two years of life before the development of language, is exceptionally long compared to other animals. This allows many opportunities for not just mother-infant bonding, but parent-child bonding also, since research shows that infants create strong bonds with their fathers from birth. The father-infant bond is not as strong as the mother-infant bond, if the mother is present. If not, the caretaker, whether it is the father or someone else such as a grandmother, creates the "neural imprint" with the infant.

This has led Wathey to explore whether the strong impression of a God that this bonding creates in the infant means that our idea of God is really that of a goddess. While there have been societies which worshiped the Mother Goddess as the supreme deity, Wathey's research suggests that the concept of God is an idealization of an all-powerful, all-loving, all-knowing creator with no sexual characteristics initially. The way this idealization is then anthropomorphized, or turned into a human, depends on the culture of the community. Christianity has created a father-god and a mother-god, by granting the mother of Jesus the position of the Mother Goddess who can intervene on our behalf with her son as God, or with God the Father. Even for those forms of Christianity which reject the idea of worshiping the Virgin Mary, Jesus can still take on feminine characteristics (gentle, loving, and compassionate) in contrast to the stern and vengeful God the Father.

One final note of importance which Wathey makes is that the potential which we carry inside of us for experiencing the presence of God, and which derives from the mother-infant neural bond, can be activated at any time in our life. Usually the trigger is some life-threatening event, at which time the individual can have a flashback to their infancy of such vividness that it stands out as one of the most important events of their life. Wathey in fact opens his book with his account of such an experience that affected him in a life-threatening situation, and he emphasizes that often the flashback is not of a specific infant experience, but of an overwhelming emotional sensation of a presence that is comforting, assuring, loving, and protective. In other words, the neural model kicks in once again, as strongly as it did when the individual was an infant.

One other way scientists can better understand the neural model, and provide more detail as to its functioning, is simply through asking adults about their experiences. It takes a lot of training for someone to learn how to ask the right questions in a proper way, so that the subject does most of the exploratory work, rather than the questioner. This is the type of training a psychologist needs in order to become a psychoanalyst. A psychiatrist needs even more training, a medical degree in fact, in order to be able to understand the neurological effect of drugs that can be prescribed to help.

To compound these demands even further, it takes an exceptional analyst to devise ways to work with children. The first such analyst was Freud's student, Melanie Klein. Her work with infants and children was seminal, and influences psychologists and psychiatrists to this day. In the next chapter, we will investigate what the Kleinian school of psychology has learned about using organized and purposeful play to uncover the inner minds of infants and children.

CHAPTER FIVE

From Infancy to Childhood

Infancy is the perpetual Messiah, which comes into the arms of fallen men, and pleads with them to return to paradise.

- Ralph Waldo Emerson

Language and the Discovery of Adult Behaviors

Sigmund Freud's contribution to psychoanalytic theory is not simply that he invented the discipline, but that he attracted so many talented students and acolytes who were able to expand the theory into many productive fields. It's true that Freud was very protective of his work and his position as the founder of psychology as a scientific and medical discipline. As a consequence he viewed any of his followers who disagreed with him as being traitors to his cause. He left behind strained relations with men who today are considered eminent theoreticians in their own right. But they were not all men, and one women who stands out as yet another psychoanalyst who came into conflict with Freud is Melanie Klein. Klein did not believe in Freud's passive approach to analysis, in which the patient was encouraged to do almost all the talking, and eventually discover things about themselves at the gentle prodding of the analyst. Besides, she wanted to explore the thinking processes of infants who had no language skills. For this, she needed to engage them on their level, and so she used puppets, puzzles and story-telling to observe and participate in the inner life of children. She firmly believed that through such activities she could penetrate the unconscious mind of the child.

Today, Klein is now credited as one of the main developers of the field of child psychology based on her decades of work with children. She died in 1960 in London, and left many followers (like Freud, she was both innovative in her theories and protective of them to the point that she would disown any followers who strayed from the true path as she saw it). The Kleinian school of child psychology, as it is called, remains prominent in Europe, but less so in the U.S. One of her principal followers in America is Dr. Chris Minnick, head of the Klein Academy in California. He maintains an active psychoanalytic practice and has also done extensive study on the latest neuroscientific advances. We will be using Dr. Minnick's work to provide an up-date on the latest thinking of the sources of God-belief in children, as determined by the Kleinian model of child psychology.

Central to Klein's theory is the concept of the "baby core" of the personality, an aspect of our personality that is developed in our infancy and childhood, and which can dictate our behavior as adults. Klein observed that the very rich fantasy life of children revealed much about the relationship of the child to their parents, and any other adults who served as principal caregivers for the child. Minnick expresses the view, consistent with Klein's observations as well, that the infant's main task is to struggle to make sense of their new world, which appears frightening and unsettled considering how helpless the infant is. He writes:

> **We are all regularly influenced, in the course of a day, by thoughts, feelings and attitudes that are emanating from unconscious structures, and states of mind, that had their origin in infancy. I find it useful in the consulting room with patients, and in my daily life outside of work, to think of we humans as having an array of positive and negative attitudes about ourselves, parents and families, and life in general. These began in infancy, in order to establish some order in the face of potential chaos.**

The infant interprets the world through its interactions with adults, most typically with the mother, father, and/or closest caregiver. Minnick notes that these interactions, in the eyes of the infant, are maintained and interpreted as "paired relationships," the infant to the mother, the infant to the father, and so forth. In establishing these relationships, the infant creates over time a sense of self – an identity of its own. It establishes a separate identity for the parent or caregiver, and in Klein terms, this is the "object" of the infant's attention and desires (Freud called this the "super-ego" or outside identity, compared to the "ego" or self-identity). For most infants, this process begins around age one, and will contribute to a sense that these God-like figures have a separate human identity from the infant, and occupy an external space as well. It is the beginning of the infant's identity as "I" and the adults who take care of them as "They." This is the same process discussed in the previous chapter by John Wathey: the development of the Parietal and Occipital Lobes allows the infant to provide physical characteristics and a separate sense of space to their parents or caregivers.

Minnick comes to these conclusions by observing the play activities of infants, but Melanie Klein was the first psychoanalyst to reach similar conclusions, only much earlier. It was one of the sources of her conflict with Sigmund Freud. Freud believed infants at birth were a blank slate – he called it the infantile "autistic phase." Klein realized from her work with infants that they had an intense inner life from birth, based almost solely on strong emotions connected to their paired relationships with parents or caregivers. Their mental life at birth, and even before birth, was anything but blank.

Klein coined the phrase "feelings as memories" to reflect the fact that these intense emotions did not simply go away. They were repressed by the infant, which would mean in

Freudian terms that they were held in the unconscious mind. They had the potential, much later in life, to reach the conscious mind under especially stressful situations, and perhaps cause problems for the individual as an adult. Again, using Freudian terms, these problems would manifest themselves as neuroses or psychoses.

For quite a long time, psychologists and psychiatrists were uncertain whether infant memories could be maintained. We will discuss in the next chapter the phenomenon of "Infantile Amnesia," a term created by Freud to describe the inability any of us has to recall memories from our infancy or early childhood. The question has always existed as to whether these infantile memories are destroyed or repressed, but modern science is revealing that the memories are not destroyed, and are indeed available for recall at unpredictable moments of stress. Dr. Minnick describes one experience he had with a young girl and her mother which gives evidence of this:

> **For example, a four year old girl, born so prematurely that she required a tracheotomy for a period of time in the neonatal ICU, was playing with me in a family therapy session. I had a rubber "monster" finger puppet with a large mouth, on my index finger. As she tentatively put her own index finger into its mouth, she recoiled as if in pain, cried "Ouch!", and simultaneously touched her neck with her other hand, and immediately looked at her mom. The mother had noticed the same thing and said to her, "You are remembering when you had a tracheotomy."**

Neither Freud nor Klein had the benefit of modern neuroscience to explain what happens in instances like this. Dr. Minnick states that humans, like other vertebrates, must be able to store early memories for their own survival, so as to avoid threatening experiences in the future. The logical place for storage of such memories is man's "lizard brain," to which we have made reference earlier, and which means storing these memories in areas like the Limbic System, or the Mid-Brain of the Brain Stem. It is the Mid-Brain which communicates with the rest of the brain, particularly the Amygdala of the Limbic System, because the Amygdala processes memories of painful, fearful, and life-threatening experiences. The sequence of near-instantaneous events affecting the four year old girl might have involved an eruption of a primitive emotional experience in the Mid-Brain, signaling a fear response in the Amygdala.

The Amygdala also is the repository of anger, and one interesting aspect of Dr. Minnick's story about the four year old girl is that when she recoiled in fear and touched her neck in memory of a painful experience with a tracheotomy (an event which occurred almost immediately after birth), she looked at her mother at the same time. She associated the painful experience with her "paired relationship." This would have been but one of many such "feelings as memories" the little girl would have ascribed to her "bad mother," the mother who did not protect her in the ICU. More typically, such "bad mother' memories are associated with a caregiver who frequently waited a very long time to respond to the cries of an infant, or who might favor another sibling over the infant. In Kleinian theory, all infants have "bad mother" memories, and in a healthy mother-infant relationship, many more "good mother" memories.

At birth, each of us has but one-fourth of the brain we will eventually develop as an adult. We rely almost entirely on our lizard brain at birth, which controls motion and sensation, and allows for emotional experiences such as fear to be generated and stored as "feelings as memories." We do not yet have a fully-functioning Frontal Lobe to process these emotions and convert them into rational thoughts. The four year old girl responded to an emotional memory which at the time of her birth could not be analyzed logically by her.

A friend of mine whom I have known for over forty years talked on occasion about his experience when he was seven years old, living with his family in the early 1940's in a small village in Eastern Europe. One day the village was overrun by Nazi troops, and he remembers having to flee the village on foot with his parents, and looking back to see the Nazis set fire to the homes and barns. This is the extent of the memory he had of the experience – a somewhat hazy but nonetheless very cruel and dangerous moment in his young life that was associated with intense fear.

Recently, it was necessary for him to undergo heart surgery to replace a valve. While he was in the operating room being prepared for the surgery, his heart stopped. In the panicky moments after his heart stopped, with nurses bustling around with a defibrillator and doctors shouting orders, he had a flashback memory of his experience as a seven year old, fleeing German soldiers. This time, it was not a hazy memory, but the replaying of a movie reel. It felt, he said, as if he was experiencing the moment all over again and he could recall the finest detail, only this time in color. The strongest impression he had from this experience was the fear that shook him so fiercely at the time. The flashback was so vivid that he had difficulty telling me about it and does not willingly talk about it with others. While he is not a particularly religious person nor a churchgoer, he insists that in those two dreadful minutes, earnest prayers to God on his part allowed his heart to start up again.

Was this "movie reel" stored in precise detail in the Mid-Brain, or perhaps the Amygdala? Were his memories lodged in his Hippocampus? A seven year old has a much more advanced brain capacity than a newborn, and this suggests that any or all of the above parts of his brain were activated in a crisis which his body recognized as existential, necessitating a reaction of extreme fear. Moreover, the Occipital Lobe was able to interact with the Hippocampus in ways that were not typical; the stressful situation caused barriers to communication within his brain to be reduced, or unused avenues of communication to open. This allowed a retrieval of the actual visual memories, and not those visual memories distorted by years of conscious reprocessing in other areas of the brain, such as the Hippocampus.

With the benefit of further research, neuroscientists may well be able to pinpoint where important childhood memories reside in the brain, and how the process of retrieval to the conscious mind works. Based on existing research, and many decades of experience psychoanalysts and psychologists have had working with patients, some observations can be made:

1. Children are not "blank slates" at birth. They have an ability at birth to experience emotions, often intensely, and these emotions are stored as memories during the first year or so in the parts of the brain that were developed at birth: the Mid-Brain, and some core elements of the Limbic System.

2. Traumatic memories of fearful or life-threatening experiences have the potential to erupt to the conscious mind of an adult under the right circumstances. Otherwise, they remain repressed in the unconscious mind, with the potential to have an ongoing effect on the adult's emotions and behaviors in negative ways.

3. Similarly, pleasant experiences are stored as well, and those experiences which are repeated constantly, such as a sense of comfort and love from a mother or other caregiver who appears all-powerful, play out in our adult lives by reinforcing the conscious impression that there exists an all-powerful figure who will protect and guide us in life.

4. Around age one, the infant begins to create paired relationships, learning to separate their identity from that of their mother, for example, who is recognized by facial and other personal characteristics, and who is also positioned externally in the outer world, physically separate from the infant.

5. Around age two, as the infant begins to learn language and progress into childhood, comprehension of the emotions of their parents and other adults begins to take place. There is a growing awareness that not all of these emotions are positive.

The Imperfect God

How do these processes of paired relationships, feelings as memories, and memory storage relate to our search for positioning of God-belief within the brain? Dr. Minnick says the following:

> **Infants universally seem to imagine that mom and dad "have everything, know everything and can do anything." Therefore, if anything goes wrong, the infant will invariably see it as their fault. They will blame themselves for an alcoholic mom, divorce, fighting, or paying little attention.**

We can parse this statement into two sections. The first half is confirmation of one of the major themes of this book: ***our image of God is derived from our infantile and childhood experiences with God-like figures, in the form of our parents or caregivers***. But very quickly in our young life, and definitely as we begin to understand language, we need to confront a paradox that cannot easily be resolved. As we begin to interpret both language and adult emotions, we discover that adults have positive and negative emotions. How could this be, if they are God-like figures and infinitely powerful in our eyes? They can't be flawed and still be God-like figures, can they?

As Dr. Minnick has discovered from dealing with these issues with patients:

> **God is a concept related to good, order, beauty, meaning, etc. God is a force or power linked in some unknown fashion to the order or meaning of the organization of the universe and its contents. God is an all-powerful and all-knowing being who created and controls the universe and its contents.**

The paradox arises because infants early on develop a basic idea of good behavior and bad behavior. Recent research conducted at Yale University by Dr. Paul Bloom has established that infants even as young as three months have an innate sense of good and bad behavior that appears to be an evolutionary inheritance common to us all. Dr. Bloom has summarized the research that was done and the conclusions obtained in his recent book, *Just Babies: The Origins of Good and Evil.* He writes the following:

> **"I think that we are finding in babies what philosophers in the Scottish Enlightenment defined as a moral sense. This is not the same as an impulse to do good and avoid doing evil. Rather, it is the capacity to make certain kinds of judgments – to distinguish between good and bad, kindness and cruelty."**
>
> **-Dr. Paul Bloom**

The infant's solution to the paradox presented by the behavior of their "all-perfect parents" is to engage in a practice that is common to all of us. Psychologists call it "splitting." Freud and his followers used the term splitting to define a "black and white" way of thinking, wherein the subject sees the world in stark, binary terms. Splitting is a defense mechanism. In infancy and early childhood, splitting allows the subject to absolve the mother, or caregiver, of any fault for their behavior which may be injurious to the subject. The benefit of splitting in this case is to provide continuous support for the image of the mother or caregiver as perfect in all respects. Without this benefit, the infant would be forced to face the insupportable: the idea that their closest protector and provider has the potential to fail them.

But there is a cost to the practice of splitting. Someone else other than the mother has to be blamed for her bad behavior. The infant has two options of assigning such blame, and will usually take on both. The first is for the infant to assume the blame itself. The infant develops a sense that they must have done something wrong to make their mother unhappy, angry, neglectful, or uncaring. These are feelings or sensitivities that the infant develops; they are not logical thoughts. At the age of two to three, when this process begins to take place, the Frontal Lobe is not fully developed enough to allow the infant or child to process their thoughts logically and rationally.

The cost of this form of psychological splitting is significant and, as we will see, life-long. As early as age two to three, we begin to develop a sense of guilt regarding our behavior, as one explanation for the new discoveries we are making about our mother or other caregivers. That discovery is a dawning realization that they are not perfect in all ways.

As adults, it seems to us preposterous that an infant or child would assume guilt for events beyond their control, but psychologists and psychiatrists have affirmed this reality time and again. I have seen this situation play out in my own experience as a parent. Our son and second-born child was born with a severe heart defect and did not survive a reparative operation done when he was nearly two years old. The event hit me and my wife like a sledgehammer, and as could be imagined, was the worst experience of our life. Our other child, our daughter, was four years old at the time. The hospital had organized a therapy program for children who had lost siblings, and she participated in the structured playtime that helped such children cope with the loss of their brother or sister. When the program was over, we were told by the therapists not to be surprised should this event reemerge in our daughter's life when she was a teenager.

Her remaining grade school and middle school years were uneventful. She was enjoying what appeared to be a normal, happy childhood. In her freshman year in high school, she started out successfully in terms of schoolwork and adjusting to the new environment. But she soon found herself among the wrong sort of friends, who skipped school frequently, hung out in local parks, and got into varying degrees of trouble, including drug use. Her grades plummeted, she became regularly insolent at home, and told us we knew nothing about how the real world worked. Only her friends out there in the "real world" knew how dangerous life could be.

We were desperately searching for solutions to this situation, when my wife remembered what the therapists at the hospital program had said about our daughter – how problems years later might erupt from the trauma she experienced in childhood. When we suggested to her that she might want to talk to a private therapist, we were surprised at how quickly she took us up on the offer. Toward the end of these sessions, we asked the therapist how they were going. The therapist said quite properly that she was unable to offer us any details, due to patient confidentiality requirements, but she did say that most of her discussion with our daughter was about the death of her brother, and how she blamed herself for his death.

When as a parent you hear news like this, you want to hug your child and tell them that of course they had nothing to do with something as tragic as the death of their brother, and they shouldn't blame themselves in the least. That is our adult, logical and rational mind coming into play, and not in a very helpful way. Our daughter was dealing with a problem which occurred when she was four years old, and at that age, a child doesn't have the fully rational mind of an adult. Any trauma she experienced at that age was going to be processed as a retrospective memory involving considerable emotional pain and stress. These were emotional experiences that at age four she could not process rationally. Our daughter's therapist understood this far better than we did, and helped our daughter work through the emotional consequences of this childhood experience, and the guilt children in such circumstances almost reflexively assume for terrible things happening earlier in their life. Happily, the therapy sessions helped our daughter to find her own bearings, and allowed her to renew her interest in schoolwork and make better judgments about her friends and her own behavior.

The Kleinian school of psychology tells us that it is not only those children who experience childhood trauma who assume personal guilt for what happened to them. We *all* have negative experiences in infancy and childhood, and these are not at all traumatic experiences, nor very frequent. Nonetheless we need to find a way to cope with them without abandoning our trust in our mother, father, or caregivers. This is the point where psychological splitting allows us as an infant or young child to create an additional paired relationship. This second paired relationship is between the infant and the bad mother who is flawed. According to Dr. Minnick:

> **The division into "parts" [paired relationships] that are generally held separately in the mind, will usually include the evacuation of the "bad" versions of self and object, into the outside world, on a semi-permanent basis, via projective processes...**

What this means is that as infants, to avoid the pain of these occasional bad experiences with our caregivers, we a) create a second paired relationship that contains the bad experiences, and b) project those out into the external world, as if they happened to someone else, and not to us.

When we talk about the "bad mother" or "bad caregiver," we are speaking in the context of a normal relationship between that person and the infant or child. Even with the best parents or caregivers in the world, the infant eventually discovers flaws in their parents' behavior which will cause pain or confusion for the infant. But what happens to us if we have less-than-ideal parents? What if our one of both of our parents are cruel? We occasionally read stories in the press about infants who are horribly abused by their parents.

The Kleinian theory says that in such cases, the infant will create an image of a cruel parent, or absent parent, or self-centered parent. This image will derive from "feelings as memories" that result from the infant's bond, and it *is* a bond, with the cruel parent. For example, they might form a bad father relationship that is characterized by one or more cruel and intolerable behaviors, *from which the infant cannot escape.* By projecting this relationship externally, on to some imaginary other infant, the behaviors can be somewhat tolerated. From the analyst's point of view, the sad consequence of having such a traumatic childhood is that later in life, as an adult, the individual will suffer from psychoses, neuroses, or other psychological illnesses precisely because the adult will have such a difficult time consciously confronting their childhood trauma, since they have spent a lifetime pretending it happened to someone else.

Depending on the circumstances, the infant could have multiple paired relationships. In the relationship just described, where the mother is loving most of the time (and tolerable all of the time), but the father is cruel and intolerable, the paired relationships would be structured this way:

- Infant to Good Mother – Infant accepts the relationship as beneficial. No guilt involved.
- Infant to Bad Mother – Infant projects the relationship externally, to an imaginary other infant. The infant assumes guilt for the bad yet tolerable behavior of the Mother.
- Infant to Cruel Father – Infant projects the relationship externally, to an imaginary other infant. The infant assumes guilt for the cruel and intolerable behavior of the Father.
- Infant to Good Father – Infant accepts the relationship as beneficial, since many abusive parents often have a charming side that rewards the infant for "good behavior." The infant may even cling to the relationship in a desperate attempt for love. No guilt involved; just a longing for affection.

The American rock singer Bruce Springsteen reported on his experience coping with his emotionally distant and verbally abusive father. As a young adult, but before he was successful as a musician, Springsteen moved out of his parent's house and found lodging of his own. After a period of time, he developed a habit of driving at night to his parent's house several times a week, parking out front, never phoning his parents or knocking on the door to inform them he was there, and simply sitting in the car for an hour before leaving. This behavior went on for such a long period of time that Springsteen realized there was something wrong with what he was doing, and he sought help from a mental health professional.

Working with a therapist, Springsteen came to the conclusion that he wanted something important from his father, and what he wanted was his father's love and approval. His reasons for not approaching the door of the house and speaking to his parents were his fear of a confrontation, and his fear – indeed his near certainty – that his father would never change his behavior toward Springsteen. Once Springsteen reached these conclusions, and came to an understanding that the love he so desperately wanted from his father was never going to be granted to him, Springsteen no longer had a need to make nocturnal visits to his family home. A Kleinian psychoanalyst might view this circumstance as one where the patient had created early in life a paired relationship with a distant father, and was attempting as an adult to upend this paired relationship and "fix it," which was something only the father could do.

How does this image of a cruel parent or caregiver impact the infantile concept of God, by which we mean a God-like figure? It is not that the infant will refuse to create an image of God – far from it. The image of a God is just as real for an abused infant as for a loved and cared-for infant. But the God of an abused infant can take on a different form. It can be a vengeful, vindictive, and cruel God, or a God to be feared. It can be a remote God, emotionally distant. Infants who have suffered emotional or physical abuse can also experience as adults an intense longing for a God-like figure of love and solicitude who was absent from their life.

Much depends as well on whether these feelings as memories are able to be processed later in childhood and then into adulthood, by being given permanent storage in the Hippocampus, and by being processed through the Frontal Lobe and other areas of the Cerebral Cortex. By the time we begin to develop language skills and enter into child-hood, we develop the capacity to interpret these memories as more than emotions. In Kleinian terms, we create an internal story to explain what the emotions mean. The stories themselves are told over and over as the feelings as memories resurface, not just through renewed contact with caregivers who have God-like authority and powers, but through what these caregivers and others teach us about God, and about religion. This is the point where culture enters the picture. If we are raised in a family of Muslims, we are given Allah as our expected God. What we think of Allah, what powers we ascribe to him, and which of his personal attributes we choose to emphasize (is he merciful, is he demanding?), are aspects of his character that are defined deep down by our infantile and childhood experiences. These characteristics of God depend on the degree to which our feelings as memories are mostly positive or negative.

The following charts show how this process works using a Kleinian model. In the first chart, an infant's overall favorable experience with a loving mother creates a loving, nurturing image of God. We describe this as a model for the Anthropomorphic (Man-created) Loving God.

The second chart is a Kleinian Child Development Model for a Cruel God.

Chart Seven: Kleinian Child Development Model of a Loving God

Kleinian Child Development Model
Example of Positive Infant Experience

	Birth to Two Years		Two Years and On	
Developmental Stage:	Infancy Pre-Verbal		Childhood Language Develops	
Event:	**Loving Mother**			
Memory Content:	"Memories as Feelings" Images, Sounds, Smells, Touch	Preliminary Explanation of Event	Explanation Ready For Long Term Processing	Internal Story Created about Event, Using Language
Lodged Where:	Mid-Brain	Hippocampus	Hippocampus	Frontal Lobe, Parietal Lobe, etc.
Hemisphere of Brain:	Right Hemisphere – External Awareness		Left Hemisphere – Internal Story Telling	
Recall Ability:	Unconscious Mind – No Recall Can Influence Thinking and Actions as Adult		Conscious Mind- Recalls as a Story Reshapes and Manipulates Story on Each Recall	
Impact:	Infant/Mother Relationship Infant as Self, Mother as "Object" to Self Infant Accepts and Internalizes "Good Event"		Ability to Form Loving, Stable Relationships	
Fantasy Created:	Template for "Loving Mother": All Loving, All-Powerful, All-Knowing, Always-Present, Creator Figure			

Model of an Anthropomorphic Loving God

CHART SEVEN

Chart Eight: Kleinian Child Development Model of a Cruel God

Kleinian Child Development Model
Example of Negative Infant Experience

	Birth to Two Years	Two Years and On
Developmental Stage:	Infancy Pre-Verbal	Childhood Language Develops
Event:	**Cruel Father** Absent, Abusive, Emotionally Distant	
Memory Content:	"Memories as Feelings" Images, Sounds Smells, Touch → Preliminary Explanation of Event	Explanation Ready For Long Term Processing → Internal Story Created about Event, Using Language
Lodged Where:	Mid-Brain → Amygdala	Hippocampus → Frontal Lobe, Parietal Lobe, etc.
Hemisphere of Brain:	Right Hemisphere – External Awareness	Left Hemisphere – Internal Story Telling
Recall Ability:	Unconscious Mind – No Recall Can Influence Thinking and Actions as Adult	Conscious Mind- Recalls as a Story Reshapes and Manipulates Story on Each Recall
Impact:	Infant/Father Relationship Infant as Self, Father as "Object" to Self Infant Blames Self for Event	Possible Lifelong Difficulty with Relationships
Fantasy Created:	Template for "Cruel Father": Distant Caregiver, Lack of Love for Infant, All-Knowing, All-Powerful, Disciplinary, Creator Figure ***Model of an Anthropomorphic Cruel God***	

CHART EIGHT

In our search for the God of our mind, the God we have created, we have observed what happens in infancy and then in early childhood, from year two and on. Sometime around year four to five, however, some remarkable things happen to our brain, and the process known as Infantile Amnesia wipes out most of the synapses, or neuronal links, that allow our unconscious memories to surface to our conscious mind. This next chapter looks at this fascinating development, because it too has a decisive impact on how we as adults view God.

CHAPTER SIX

Confronting Death

Men come and they go and they trot and they dance,
and never a word about death. All well and good.
Yet when death does come - to them, their wives, their
children, their friends - catching them unawares and
unprepared, then what storms of passion overwhelm
them, what cries, what fury, what despair!

-Michel De Montaigne

The Fear of Death

Does anyone you know go about their daily lives obsessed with death? For most of us, death comes to mind only when it is brought to our attention. We read about a terrible plane crash with multiple fatalities; a follow-up story relates how some people changed their travel plans at the last minute and avoided that flight. An earthquake strikes in some place far away, and thousands die. A celebrity dies unexpectedly of a drug overdose. All of these reminders cause us to think momentarily about our death, and how fortunate we are to have avoided these tragedies. Then we set these thoughts aside and go about our daily routine.

It is only when we ourselves confront death that something deeper and more meaningful happens to us. That lump you suddenly discover in your breast – what is that? Is it cancer? Are you going to die? These thoughts occur in such a rapid sequence that we think they are one instantaneous thought, accompanied by something just as significant: *fear*. It is the same, sudden fear we have when a car pulls into our traffic lane so quickly that we must slam on the brakes to avoid a collision. "I could have died just then!" we say. We are angry at the other driver, and we may shout out our anger, but underneath it all, we were momentarily *afraid*. The panic we felt was completely out of our control.

Neuroscientists can tell us what happens in our primitive brain at such moments, and how our Amygdala (and even our Mid-Brain) sends a surge of fear through our body, causing our muscles to react with the stress. We start panicking and sweating if we really think we are going to die from cancer; we grip the steering wheel tightly if we have to make a split-second decision to avoid an accident. This is our brain reacting in survival mode, using fear as a means to save our life, putting us on the highest alert, activating all of our senses when danger arises from nowhere. These are the same reactions our earliest ancestors had on the plains of Africa when they detected a dangerous prey that was stalking them. These are our fight or flight reflexes, brought to us by millions of years of conditioning, hard-wired into our brain in such a way *that we are not in control of the fear* that motivates us to take action.

The fact that fear of death is something we cannot control tells us something very important about ourselves: our brain and our body are built to fear the possibility of death, and to do everything possible to avoid death. Fear of death is a primal instinct, but what is coming to be understood more conclusively, is that *fear of death is **the** primal instinct.* If there is one single thing that motivates all humans to respond in the same way, it is the fear of death.

Perhaps it is just as well, then, that the fear of death is something that we need not face at the moment of our birth. It takes years for the average child to come to a full realization that they are going to die, and the range around which this can happen is quite broad. Child development specialists have noted that this realization can occur as early as age three, and as late as age ten. What makes it especially difficult to study the fear of death in children is that few children remember that one, horrifying moment when the truth of their mortality hit them. There are several reasons why most children cannot recall when this happened, and we will talk about these reasons shortly.

Let us just say that it is a rare person who remembers this moment of revelation, or series of moments if the knowledge came gradually. One such person was the Jesuit priest, Pierre Teilhard de Chardin, who had a specific and very painful adult memory of the particular moment around 1887 when he learned he was mortal:

> **A memory? My very first! I was five or six. My mother had snipped a few of my curls. I picked one up and held it close to the fire. The hair was burnt up in a fraction of a second. A terrible grief assailed me; I had learnt that I was perishable... What used to grieve me when I was a child? This insecurity of things. And what used I to love? My genie of iron! With a plow hitch I believed myself, at seven years, rich with a treasure incorruptible, everlasting. And then it turned out that what I possessed was just a bit of iron that rusted. At this discovery I threw myself on the lawn and shed the bitterest tears of my existence!**
>
> **-Father Pierre Teilhard de Chardin**

Teilhard de Chardin was recognized by his parents as being both highly sensitive and observant at an early age, qualities which he developed later as an adult into a deep spiritual and mystical attachment to Christ, combined with an extremely practical, and literally down-to-earth obsession with paleontology and the origins of early man. Perhaps we should not be surprised that such a person had as his earliest memory the most painful emotional experience of his entire life: the discovery of his inevitable death.

Even granting that Teilhard de Chardin was an unusually sensitive child, why would a child as young as seven interpret this knowledge of death as a devastating, existential revelation? The answer is biological in nature. We humans are in the vertebrate family, which consists of animals with spines. All higher-level vertebrates have an Amygdala, which means that even the lowly lizard has the same fight or flight response that we have. The difference is that we are the only animal that can contemplate our death in advance of the event. All other animals, in the words of Arnold Schopenhauer, "hear about death for the first time when they die." We are burdened with a Cerebral Cortex and an elaborate memory system that allows us to imagine what our death would be like. We understand the enormity of what is at stake if we were to die. There is nothing we hold so dear as our life.

Just as we hold life as that which is most dear to us, so we hold death as that which is most fearful to us. This reflects the essential duality of our human character, which was first described by the Danish philosopher Søren Kierkegaard. He characterized man as having simultaneously the qualities of infinitude and finitude. Infinitude is the ability to imagine a universe without limits, to believe in a place called heaven which exists for all eternity, and to contemplate ourselves residing ultimately in heaven and worshiping a God who is omnipotent, omniscient, omnipresent, all-loving, and in every other way without limitation. At the same time, we are constrained by the finite, cruel realities of our body and material existence, and the fate that we all await, which is our decline and eventual death. In his classic 1974 book *The Denial of Death*, which explores the psychology resulting from this duality, Ernest Becker wrote:

> This is the paradox: he [man] is out of nature and hopelessly in it; he is dual, up in the stars and yet housed in a heart-pumping, breath-gasping body that once belonged to a fish and still carries the gill-marks to prove it. His body is a material fleshy casing that is alien to him in many ways— the strangest and most repugnant way being that it aches and bleeds and will decay and die. Man is literally split in two: he has an awareness of his own splendid uniqueness in that he sticks out of nature with a towering majesty, and yet he goes back into the ground a few feet in order blindly and dumbly to rot and disappear forever. It is a terrifying dilemma to be in and to have to live with.

Becker went so far as to call the fear of death, "the terror," saying, "This is the terror: to have emerged from nothing, to have a name, consciousness of self, deep inner feelings, an excruciating inner yearning for life and self-expression— and with all this yet to die."

Shortly after publishing *The Denial of Death*, which won the Pulitzer Prize for non-fiction and was instantly recognized as a significant contribution to the field of psychology, Becker was diagnosed with terminal cancer. Did he know at the time of his death the importance of *The Denial of Death*? The book has led to a whole new branch of social psychology, termed Terror Management Theory (TMT), which applies Becker's considerable insights to many different areas of human behavior, of which God-belief and religion are just one area of interest within the broader TMT field.

It would seem on the surface that the fear of death cannot be as significant as Becker claimed in *The Denial of Death,* and he did in fact devote some time in the book to those psychologists and psychoanalysts who thought that the fear of death existed but played a minor role in human behavior. Becker came down firmly on the side of those who felt otherwise, particularly psychoanalysts, who had the advantage of treating hundreds if not thousands of patients whose psychological problems often found their roots in a deep-seated anxiety over mortality. What convinced Becker of the primacy of the fear of death in our lives were the many ways in which we act on a daily basis as if death didn't exist. These are deliberate efforts to ignore death, and overcome it as well. In different terms, we attempt to deny death, and we seek not-death.

Psychologists and psychoanalysts have also observed that the fear of one's individual death is part of a more complex anxiety – *fear of the external world, or Nature.* As infants, our very helplessness plays an important role in creating an emotional template that the world "out there" is unpredictable and dangerous. As infants we understand how little control we exercise over the world, and our development into childhood and adulthood is to a large degree an effort to exercise more control over the world, and thus over our lives. Inevitably, however, we develop a conviction that we will never have the control we desire over Nature.

It is true that modern man is no longer as closely associated with Nature as were our ancestors. Europeans as recently as the 17th century were in regular fear of Nature; wolves for example were a common predator which men feared to such an extent that they featured prominently as representations of evil in fairy tales. But let us not congratulate ourselves that we have in modern times escaped the fear of death that Nature constantly presents. For modern man, Nature stalks us not through carnivores, but through the microscopic. Think how quickly an entire globe of seven billion people reacted with panic at the recent outbreak of the Ebola virus in Africa. It was as if our entire species had a collective Amygdala which was activated at the mere whisper of the word "Ebola."

It is true that there are many occasions when we can enjoy the majesty of Nature, and credit God for providing us the beauty that exists in the world. But as with anything to do with God, he is credited only with the qualities of goodness that are sometimes evident in Nature. However beautiful the world might be, at a deeper level we know that Nature is in control ultimately of our existence. Our death is intimately associated with Nature, because all animate things die – it is how Nature is constructed, and the corrupt, material aspect of our being is part of Nature. This is why we blame the existence of death on Nature (as represented by the Grim Reaper, e.g.), or Satan, or some other aspect of our lives that is the opposite of God. He represents for us eternal life, at the price of course of our obedience to him. God is, as we have said before, the default belief for mankind, and he is therefore our default avenue to not-death.

Psychological Repression

There is simply no way we can live our lives in constant fear of death. We know the fear is real; we've all experienced some brief moments when have been forced to confront the reality of our earthly mortality. None of us can imagine the pain if we had to experience such fear with any frequency. Our species would not have survived if we had been burdened by the incessant distress of pondering our ultimate demise. From an evolutionary perspective, the human species had to find a way around this problem many eons ago, and psychologists define this solution as "psychological repression."

It is tempting to define psychological repression in loose terms, as if it were nothing more than an attempt to forget about unhappy events or potentialities. From our personal experience of psychological repression, we tend to think it is the practice of imagining our death occurring far into the future. It is much more than that. Says Becker:

> But repression is not a magical word for winning arguments; it is a real phenomenon, and we have been able to study many of its workings. This study gives it legitimacy as a scientific concept and makes it a more-or-less dependable ally in our argument. For one thing, there is a growing body of research trying to get at the consciousness of death denied by repression that uses psychological tests such as measuring galvanic skin responses; it strongly suggests that underneath the most bland exterior lurks the universal anxiety, the "worm at the core."

It was the American psychologist William James who first described the knowledge of death as "the worm at the core" of the human character. James had a sense that the knowledge of death was something that was at the center of childhood insecurities, but it was with the benefit of nearly a century of thought and research on the matter that allowed Ernest Becker to describe the importance of the fear of death to childhood development:

> An increasing number of careful studies on how the actual fear of death develops in the child agree fairly well that the child has no knowledge of death until about the age of three to five. Only gradually does he recognize that there is a thing called death that takes some people away forever; very reluctantly he comes to admit that it sooner or later takes everyone away, but this gradual realization of the inevitability of death can take up until the ninth or tenth year.

Somewhere in this process of gradual confrontation with death, we do more than realize that death "sooner or later takes everyone away." We learn that *death will inevitably take us away.* This is not an understanding we can achieve as an infant, but by the time we are two, the growth in our cerebral capacities is explosive. Not for nothing does the scientific community call this the "exuberant period." Millions of new synapses connecting many areas of the brain are being formed hourly, in part so we can learn language, and in part so that we can form and utilize the complex thinking necessary to navigate the real world.

It is this expansion in our Cerebral Cortex that allows us as children to comprehend death, and confront the terror of our own death. While this can occur as early as age three, it can take many years for most children to fully complete the journey of understanding, which begins with the comprehension that a squashed bug or animal does not return to life, whatever cartoons may suggest. From here, we learn that the person lying in a coffin at a wake is a human being who will never return to life. Progressively it occurs to us that our parents or caregivers will inevitably be that person lying in a coffin, at which point we will truly be an orphan in the world. Then at last we come face-to-face with the horror that *we will inevitably be that person lying in a coffin.*

From the studies that have been done on psychological repression, it is now evident that the fear exists- it is merely pushed down below the surface. Becker mentioned research using "galvanic skin responses," a technique which measures reactions in the sweat glands to stressful situations that do not appear to bother test subjects in a conscious way. The test subject may be shown slides picturing dangerous or fatal occurrences, and asked to rate their reaction to these pictures. The test subjects will show only a moderate level of concern, claiming that the events are not occurring to people they know or love, and so they are not difficult to tolerate. Their skin responses betray a more serious reaction, pointing to an immediate and substantial unconscious fear response the minute the image is perceived.

Psychological repression does precisely what its name suggests: it pushes down to the unconscious mind the fear reaction, preventing the conscious mind from having to deal with the stress. Unfortunately, the process can often merely postpone the need for the conscious mind to deal with the fear. Any psychoanalyst, psychiatrist, or psychologist will relate how real the process of psychological repression is, having helped patients whose neuroses, psychoses, or other mental problems are discovered to result from some childhood trauma that could not be easily remembered. Modern health professionals also have more diagnostic tools at their disposal than did Freud, and they know much more about the brain from advances in neuroscience. They know that as therapists, they must be careful about the "power of suggestion," whereby the patient takes a suggestion from the therapist and creates a false memory, believing that the suggested event occurred when it did not. Moreover, many mental problems are not the result of psychological repression, but of damage done genetically or otherwise to the functions of the brain. These problems can be better treated with cognitive behavioral therapy or with pharmaceuticals.

Infantile Amnesia

There is a related phenomenon the mind uses to deal with the fear of death that works its results over a long period of time. This is the phenomenon of Infantile Amnesia. There is at work during childhood a massive restructuring of our brain that buries deep into our unconscious mind not just unpleasant infant and childhood memories, *but virtually all infant and childhood memories*. How often have you played a game with some other adult, titled "What is your earliest memory?" We all have played this game, recalling our first memory of a childhood event at age four or five, and for some people as early as age two. The earliest memory is often something traumatic, such as a medical emergency involving a hospital stay or surgery, but the phenomenon is absolutely universal, and inherent in our species. You will not meet with someone who, by some freak of nature, remembers all their childhood experiences. We are each of us hard-wired to forget our infancy and childhood at the conscious level.

Psychologists of the 19th century were the first to write about this phenomenon, but Sigmund Freud was the first to give it a name – *Infantile Amnesia*. Freud developed a theory regarding Infantile Amnesia which described it as a survival mechanism for our species, but he went off track by suggesting the cause of Infantile Amnesia was trauma, brought on by the infant's anxieties over psychosexual dynamics, such as the desire to kill the father, or the fear of castration. Such theories have now been discredited, and to the extent there is a psychological aspect associated with Infantile Amnesia, it is likely to be *fear of death*. Your inability to remember exactly when you first understood the terror of your own inevitable death can be credited not just to psychological repression, but to the massive amount of memory repression that occurs under Infantile Amnesia.

The exact process by which Infantile Amnesia unfolds is not clear. It is known that from the moment of birth, and indeed before birth, the brain is expanding rapidly, but it takes over twenty years for any of us to develop an adult brain. Teenagers, for example, lack a truly adult sense of risk vs. reward, because their brains are not fully formed. They often engage in risky behaviors as teenagers that begin to seem to them to be highly inappropriate when they reach their twenties.

At birth, the Limbic System is incompletely formed, but it is capable of storing memories, as is the Mid-Brain in the Brain Stem, according to some research. These are emotional memories, which cannot be cognitively interpreted by the infant because the Cerebral Cortex operates only on a rudimentary level. The emotion may be something as basic as the happiness associated with learning a new skill, which gives the infant more control over their life. At less than six months old, infants have been shown to be able to repeat tasks for a period of several days before they lose the memory of a taught behavior (such as manipulating a toy). This period of memory and learning lengthens throughout infancy, until it reaches a two-week retrieval capability by the time language starts to develop. These multi-week memories may be stored in the Mid-Brain, the Limbic System, or even the developing Cerebral Cortex – neuroscience has not yet developed a detailed explanation of this process. Certainly, though, the infant brain as it approaches childhood at age two is creating millions of *synapses*, which are the connections between neurons that allow electrical and chemical messages to be sent from one part of the brain to another. Synapses are also presumed by neuroscientists to retain memories in some fashion, though some scientists dispute this. In any event, synaptic development in the infant brain is not simply the act of adding capacity. It is building the highly complex communications system that turns a primitive brain into a unique, living computer.

By the time a child enters pre-school age, at around ages two to five, they have an effective short-term memory which will continue to expand during their childhood. It is around age four, however, when something absolutely remarkable occurs: *the brain begins to turn off millions of synapses relating to childhood memories*. With the synapses deactivated, the neuronal paths leading from memory storage areas to the Cerebral Cortex prevent these memories from being readily retrieved and processed. By age ten, most of us have few surface memories of our childhood, and as adults, we find our childhood and infancy years to be shrouded in a very dark cloud.

Some neuroscientists argue that the brain deactivates childhood synapses in order to make room for the rapid synaptic growth necessary for the remaining seventy-five percent of brain development to take place from childhood to adulthood. The brain needs substantial synaptic growth merely to be able to process language, according to some theories explaining Infantile Amnesia. This may be as valid an argument as the psychoanalytical belief that repression of painful experiences in infancy and childhood are the causes of Infantile Amnesia.

Other research has shown that Infantile Amnesia occurs around the same time as a child develops self-awareness. Self-awareness is described by psychologists not simply as conscious knowledge of one's existence, but knowledge of one's emotions, of one's thoughts and thought processes, and of how one's behavior affects others, in good and bad ways. Beginning around ages three to four, children develop a sense of the past and future, and can distinguish items that belong to them and no one else. Gradually, children become more aware of their own emotions, and can discuss how they feel.

Neuroscientists believe a number of areas of the Cerebral Cortex are involved in self-awareness, such as the junction between the Parietal Lobe and the Temporal Lobes, as well as the Insular Cortex. Tests done with subjects who have experienced significant damage to these areas show that some of the subjects maintain their self-awareness, indicating that self-awareness may be the consequence of the chemical and electrical messaging structures that connect these areas of the brains, and which allow self-awareness to function even when component areas of the brain are damaged. My own speculation is that as these self-awareness structures develop in a child, the brain may find it necessary to eliminate synapses linking infant and early childhood memories to those areas of the brain that would normally allow such memories to be retrieved and processed. The development of self-awareness might be a process that is concomitant with Infantile Amnesia, and linked in some way, since both of these processes occur at onset from early childhood, and extend through later childhood.

For our purposes in searching for neurological sources for God-belief, the important aspect of Infantile Amnesia is that while synaptic connections are disabled in the process, the underlying memories are not destroyed. This is becoming evident from recent studies done on infant and childhood memories. In a study conducted by Kjell Morton Stormark of Norway, and published in 2004, twelve preschool students aged around two to four years old were shown slides of the faces of children unknown to them, children currently classmates of theirs, and children who had been classmates when they attended an earlier preschool. The children were tested on their stated recognition of these faces, and all of them were successful in indicating their current classmates, or denying knowing those of the strangers, but only three said they remembered the faces of earlier classmates. In addition to an overt acknowledgement response, the researchers obtained biometric responses that are known to provide a more accurate measure of emotions. These biometric responses included heart deceleration markers which would indicate some recognition of the earlier classmates if the heart rate slowed, and skin conductance responses showing elevated sweat gland production when an earlier classmate was recognized. These biometric tests showed the subjects *did indeed* recognize nearly all their previous classmates when compared to the faces of strangers. The recognition, however, was on an unconscious level, and exactly where the memories are stored is a subject for further study. These visual memories may be stored in the Occipital Lobe, or perhaps in a more primitive part of the brain, such as the Mid-Brain.

In a more recent study, published in 2014, researchers and neuroscientists at McGill University in Montreal, and the Montreal Neurological Institute, reported that the language infants are exposed to within their first year of life has a "unique and lasting influence" on their brain organization. The study focused on a selection of Quebec children who were organized into three groups: those who spoke only French from birth; those who grew up in bilingual households and spoke French and Chinese; and those who were Chinese from birth and were adopted before they were age three, by a Quebec family. The children from the last group were therefore exposed to Chinese language early in life, but spoke no Chinese now and did not understand the language. The test subjects were asked to perform tasks while listening to Chinese language sounds, and undergoing an *f*MRI review of their brain functions at the same time. The *f*MRI results for the first group showed that during the tasks, the areas of the brain normally associated with speech were activated, chiefly in the Left Hemisphere Frontal Lobe, and the Insular Lobe. The other two groups, however, had additional areas of the brain that were activated, in the Right Hemisphere and other areas of the Cerebral Cortex typically in use among bilingual speakers. This indicated that aural linguistic experiences of infants were stored in some fashion in a newborn's brain. According to Lara Pierce, the lead author of the study:

> **What we discovered when we tested the children who had been adopted into French-language families and no longer spoke Chinese, was that, like children who were bilingual, the areas of the brain known to be involved in working memory and general attention were activated when they were asked to perform tests involving language."**
>
> **-"Past Experience Shapes Ongoing Neural Patterns for Language,"**
> **Nature Communications, 2014**

If our brain is capable of storing infantile or pre-school memories of the faces of our classmates, as well as language sounds, it is certainly capable of storing memories of an all-powerful, all-knowing, all-loving, always-present, and all-providing caregiver who is essential to our survival from infancy through childhood. These memories are most often those of one or more of our parents, but any loving and caring individual will do. These must be among our most powerful and persistently-reinforced memories as infants and children. In this opening section of the book, we've provided many different scientific explanations for the process at work, suggesting that the essential mother-infant bond creates our first experience with the comforting presence of a God-like caregiver, and that the infant brain stores these experiences in the form of "feelings as memories." We've come to understand that we also carry within us a model for a God who is not always benevolent, but can be demanding and vengeful, using his omnipotence for ugly purposes. We've learned many interesting facts about where in the brain these memories are stored, how they become processed as circumstances we can analyze conceptually, and how they have the power to affect our behavior as adults.

We've also begun to explore how the fear of death can serve to repress our childhood memories, but what we want to understand better is to what extent the fear of death motivates coping behaviors we use as adults. The next chapter looks at two such coping behaviors, which supplement psychological repression and Infantile Amnesia. These behaviors include transference, and the creation of a hero project for us to follow.

CHAPTER SEVEN

Transference and Our Personal Hero Project

He did what heroes do after their work is accomplished; he died.

- Leo Tolstoy

The Power of Transference

Psychoanalysts have long known about the importance of transference in a therapeutic setting. Freud was the first to write about this process and analyze it, based upon thousands of hours of experience with patients. He realized that as a therapist, he needed to create a bond of trust with his patients, so that each patient felt liberated to say things that would otherwise not be said in public, or even privately to themselves. The trust was predicated upon a priest-confessor model, whereby the therapist promises to keep any discussions with their patient confidential. This remains today a fundamental element of the therapist-patient relationship, to the point that modern legal systems recognize that a therapist cannot be obligated in court to break the promise of confidentiality.

When a bond of confidentiality and trust is created between the therapist and the patient, the patient feels free to *transfer* to the therapist the patient's anxieties and fears. The therapist is perceived as taking on the burden of these anxieties on behalf of the patient, liberating the patient to explore more readily the reasons for these anxieties and fears. The patient is also transferring great power to the therapist, which can pose problems. The patient may begin to worship the therapist as a savior-figure, something Freud experienced in his own psychoanalytic practice. This was one of the circumstances which prompted Freud to consider that God-belief and religious practices may serve as a similar therapeutic function for many people looking for a savior and for someone willing to take on their anxieties and fears. Many people, in fact, are religious in belief and practice all of their life, deriving benefits from their religious faith that are not dissimilar from the benefits a patient seeks from their psychoanalyst. The problem for the psychoanalyst is that he or she cannot of course serve in the role of savior forever; at some point the patient must wean themselves away from a reliance on their therapist as a source of comfort and as a figure who will save them from their problems. A good part of the training of any mental health professional involves teaching them how to help their patients end the process of transference so that they function in their life on their own.

But even the healthiest individuals engage in transference. It begins with our dependence on our parents or caregivers. Long after we have embarked on our adult careers, not just via employment, but through adult relationships and the creation of a family, we still find ourselves relying on our parents for advice and comfort. Society jokes about the "returning child," meaning the adult who is out of work or suffering through a divorce, and finds as a solution to their problems that it is necessary to move back in with their parents. What is no joke is the degree to which society ultimately accepts these situations. It is perfectly natural for a forty-year old to move in with their parents because to whom else would they turn for such an imposition? The forty year old may privately or even publicly bemoan their failure to conquer life on their own, but deep down, they do not question the naturalness of relying on their parents for support in times of stress. We have each of us, to some degree or another, transferred to our parents our anxieties regarding life.

It is only at the end of *their* life, when our parents are weak and in need of help, that we as their children take care of them, not just as a filial obligation, nor simply as a means of repaying them for raising us, but in appreciation for their being our refuge and source of security throughout our lives, as long as they are alive. This is why the loss of a parent is so painful for most people, and why the loss of both parents establishes an abiding sense of abandonment in us. To whom do we now turn in a crisis if both our parents are dead? Are we prepared to truly be on our own?

Of course it is important to have a spouse or a close relationship to rely upon once our parents are dead, but that relationship is one of equals, lacking the childhood imprints that create our lifelong dependency on our parents. We cannot as readily transfer our anxieties and life-fears to our spouse as we can to our parents, because in a relationship of equals we must serve as an equal source of support for our spouse. In our relationship with our parents, there is no such sense of equality. We are always their child; we are always dependent on them to some degree.

As we have seen in earlier chapters, if we did not have a happy childhood, supported by the love of our parents or caregivers, we still have a need to transfer our fears to some person of power. To the extent an abused or abandoned child suffers later in life from relationship problems, they often involve an overreliance on people they perceive as strong and protective.

If as adults we had a happy childhood, we still need our saviors. We judge our managers at work by the extent to which they serve our interests. Are they protective of us, defending us in front of senior management, taking an interest in our careers, and ensuring that we are properly compensated? It is a mantra of modern business that the employee is responsible for their own careers, and for promoting themselves accordingly through networking and other practices. It is *the reality* of modern business that the manager has tremendous power over the employee, and given that power dynamic, what we as workers instinctively do is transfer to the manager our anxieties over the insecurity of our situation.

We seek out societal heroes everywhere. We identify with a local sports team, and feel genuine emotional sorrow when the team is not victorious, just as we interpret their victories to be our victories. In the parlance of the sports world, we as fans *worship* players who are notably successful in sports. These are god-like characters in our view, and of course the danger of treating them like gods is that our disappointment will be even more severe when they fail us on the field of play, or in their personal life.

So too do we transfer to political figures our hopes and anxieties. This is why politicians put so much effort into building up their personal image as that of a strong individual who can accomplish nearly-miraculous things, like providing jobs for most people. It is a well-known phenomenon that in times of economic and social stress, voters seek out authoritarian politicians, men (almost always men) who project self-confidence, who identify the enemies of the people, and who promise to work miracles by vanquishing enemies, and restoring greatness or grandeur. It is the nature of authoritarians that they divide the people, by emphasizing tribal identities such as race, educational level, economic class, and religious affiliation. This enhances the prospects for transference: it is easier for someone to transfer their trust to a politician of the same religion, for example.

There is another aspect of transference that operates on the social level yet mirrors the workings of transference in a psychoanalyst-patient setting. Just as the psychoanalyst provides the patient the ability to express whatever is on their mind in a safe setting, with no repercussions, so too does the authoritarian politician provide a blanket exemption to his followers over any pernicious behavior they undertake in his name. Ernest Becker said this about the authoritarian figure:

> **Furthermore, he [the authoritarian] makes possible a new experience, the expression of forbidden impulses, secret wishes, and fantasies. In group behavior anything goes because the leader okays it. It is like being an omnipotent infant again, encouraged by the parent to indulge oneself plentifully, or like being in psychoanalytic therapy where the analyst doesn't censure you for anything you feel or think. In the group each man seems an omnipotent hero who can give full vent to his appetites under the approving eye of the father. And so we understand the terrifying sadism of group activity.**

It is this "terrifying sadism" which leads first to the expression of forbidden thoughts, of racial epithets that in the past were considered in bad taste, of comments which turn "the others" into sub-humans, or of fears expressed openly about someone else's religion. From there it is a relatively easy step to violence, especially if the leader sanctions or openly encourages violence. One of the remarkable aspects of the prosecution of Nazis during the Nuremberg Trials, and as well over further prosecutions of men like Adolf Eichmann, was the uniform propensity of the Nazis to claim they were only following orders. The Nuremberg judges did not allow this defense to justify the atrocities committed by the Nazis, but it was left to the social psychologists to explain why this defense seemed so logical and natural to the Nazis themselves. They had been swept into, and they had bought into, a system wherein they had transferred the moral responsibilities for the crimes they committed to their hero – the Fürher. He had already absolved them of their actions and even of their thoughts. The transference was so complete, from the infantilized Nazi follower to a God-like figure in the form of Hitler, that the Nuremberg defendants seemed honestly puzzled as to why anyone should think they personally were guilty of a crime.

It was also remarked by observers at the time that in the pre-war years, all of Germany seemed engrossed in the transference process, investing their faith in Adolf Hitler as a savior. Hitler had a keen sense of the theatrical, and one of his tricks was to arrive at Nazi rallies by plane, as if descending like a god from the skies. Even thirty or forty years after the war, Germans who lived through that time found it hard to put into words what it was like to be part of a national mania, but there were those who continued to refer to the Nazi era as a time of glory that they did not really regret.

There is ample evidence that transference is not limited to our infant or childhood years. It is an adult defense we all use to allay our innate fears about the world in which we must operate. It allows us, as adults, to seek God amongst us, and to revert to our childlike ways when we truly were in the presence of God-like figures. It opens the door to one additional defense mechanism we use to deal with our anxieties, and that is the establishment of our personal hero project. We will explore next the use of the personal hero concept as yet another of our attempts at achieving not-death.

The Hero Project

It was in 1949, when *The Hero of a Thousand Faces* was published, that someone was first able to tie together our human desire for not-death, as told to ourselves in timeless myths, with the psychoanalytic work of men like Sigmund Freud, Carl Jung, and Otto Rank. The author of *The Hero of a Thousand Faces* was Joseph Campbell, a scholar who invented the field of comparative mythology. By studying human myths in dozens of cultures, Campbell noticed striking similarities with these stories, whether they were

200-year-old fairy tales from Germany, or episodes from Greek or ancient Egyptian mythology. From these similarities Campbell constructed a template that was universally applicable to all myths, though most myths consisted of only parts of the template. In this template the myth always features a hero, most often a male because it is men who create most myths. The hero goes on a journey and encounters the supernatural, which represents forces or beings who have power over death. The hero is tested, and survives the ordeal, after which he must return to earth bringing gifts for mankind which can be used by all of us in our mundane, natural existence.

In Campbell's template, there were three stages to the hero's journey: Departure, Initiation, and Return.

Departure: The hero is introduced as an innocent, who is somewhat out of place in his own culture. A circumstance arises which sends him on a quest or journey – the desire to join the Greek army to avenge the abduction of Helen by the Trojans, to cite just one example from Odysseus' journey. A more modern example would be the charge imposed on innocent and insignificant Luke Skywalker to take up the daunting challenge of fighting the Empire in order to save the world. Usually a supernatural being urges the innocent "everyman" to undertake the quest, as Obi Wan Kenobi did in the _Star Wars_ trilogy, and this mentor gives the hero tools or devices with supernatural powers to help in the quest. Once on the journey, the hero reaches the "point of no return," when he must cross over into the supernatural world to undergo an ordeal.

Initiation: Once in the land of the supernatural, the hero comes to understand the dangers he now faces. He must summon up the courage to face these dangers, and train himself in preparation for the battle. He will have allies in the supernatural world, supernatural beings who want him to succeed and will help him in critical moments. While it may appear that the hero's success is entirely dependent on these heavenly interventions, the hero is ultimately the determining force of the outcome of the battle, because only the hero can summon up the courage to face demons. In Mozart's 1791 opera _The Magic Flute_, the hero Tamino uses the magic flute to pass through the trials by fire and water, but it is only through his volition and personal courage that he first encounters the trials, and puts his trust in the supernatural powers of the flute to help him. The ordeal, in other words, is _always a battle the hero has within himself._

Return: With success in his battle within himself to find courage, the hero is rewarded by the gods or supernatural beings, and is tempted to remain with them in the supernatural realm. Ultimately, the hero decides to recross the boundary back into the natural world, where he is celebrated for his accomplishments and for the gifts which he has brought back to benefit mankind. These gifts can be as simple as bringing the benefits of fire to mankind (Prometheus in Greek mythology), or as magnificent as saving Middle Earth from the forces of evil, as Frodo did when he destroyed the power of the Dark Lord Sauron, in Tolkien's _The Lord of the Rings_ trilogy. Upon his return to earth, the hero undertakes a new life as a teacher, as happened in Andy Weir's 2011 novel _The Martian_, made into a popular movie in 2015. The hero in this book, Mark Watney, is an astronaut stranded on Mars, who with the help of gifts from the gods of science (botany, aeronautics, etc.) and from his fellow-astronauts, returns to earth and takes up the role as a teacher of survival methods to other astronauts. The hero ultimately becomes a mentor, so that others can undertake with his help their own heroic quests, and the cycle of heroism can continue.

The Appeal of the Hero Project

Joseph Campbell drew his template of the hero's journey as a circle, to show how the cycle is itself a process of rebirth, wherein one man makes the journey and returns to help others do the same. Campbell's tremendous erudition, including his knowledge of many ancient and modern languages, helped him decipher different myths and find the similarities. In this respect he was like a contemporary scholar, J.R.R. Tolkien, who was fascinated by Norse, Celtic, and other mythologies and who patterned *The Lord of the Rings* books on the hero's journey as he understood it from these ancient myths. Tolkien, in a sense, discovered what Campbell was coming to understand in his comparative work on mythology, but Tolkien's personal heroic quest took him into the world of fiction, where he created an immortal work of literature. Campbell created an enduring scholarly classic, in which he also explored *why* so many human cultures over mankind's history developed mythologies with the same hero's quest as the inspiration.

To answer this question, Campbell turned to the work of great men who had made advances in the field of psychology. Freud was fascinated by mythology, and related his psychoanalytic theories to mythology in his books *Totem and Taboo*, and *Moses and Monotheism*. Carl Jung developed these ideas further with his exploration of the collective unconscious, and the propensity of man to create archetypes, which can be heroes across many cultures (Loki the Norse god of mischief, is an example of the trickster archetype, which would include The Fox in the *Br'er Rabbit* stories, or Bugs Bunny in his battles with Elmer Fudd). Archetypes need not be people or animals. Jung noticed the universality of the myth of a global flood, sent by God or the gods to punish mankind, but who save one man to undergo a quest of courage so that mankind would not perish from the face of the earth. The relationship between Campbell's work and psychology has in its own way come full circle: it is now psychology which is borrowing from Campbell. A whole field of cognitive therapy is today being used to help patients understand the value of identifying their own heroic quest – their own *hero project*, as it is called, the theory being that we all have a hero project which we desire to express.

Ernest Becker claimed that the hero project was a critical element in our search for not-death:

> **When we appreciate how natural it is for man to strive to be a hero, how deeply it goes in his evolutionary and organismic constitution, how openly he shows it as a child, then it is all the more curious how ignorant most of us are, consciously, of what we really want and need.**

This was, in Becker's view, the paradox of the human condition: each of us seeks to be a hero, but we are largely unaware of this compelling desire, and unwilling to confront our need for heroism. Our ignorance of our own hero project stems from our inability or refusal to confront our fears of death. Because of this, and because modern society could not function if everyone acted upon a grandiloquent hero project like those undertaken by Odysseus or Frodo Baggins (the opportunities of which are quite minimal), society fashions for most of us ready-made hero projects. These are projects which benefit civilization, reward us with the compensation necessary to live a satisfactory life, and give us the emotional satisfaction that we are participants in a larger, grander scheme of heroism. Even something as basic as building a new computer application for an office can be viewed as a hero project for the team members who participate in the project.

In his characteristic, compelling prose, Ernest Becker spelled out how society pulls together many types of hero projects into an overriding structure that allows complex modern civilizations to prosper and regenerate themselves. He called this structure a "symbolic action system."

> **The fact is that this is what society is and always has been: a symbolic action system, a structure of statuses and roles, customs and rules for behavior, designed to serve as a vehicle for earthly heroism. Each script is somewhat unique, each culture has a different hero system. What the anthropologists call "cultural relativity" is thus really the relativity of hero-systems the world over. But each cultural system is a dramatization of earthly heroics; each system cuts out roles for performances of various degrees of heroism: from the "high" heroism of a Churchill, a Mao, or a Buddha, to the "low" heroism of the coal miner, the peasant, the simple priest; the plain, everyday, earthy heroism wrought by gnarled working hands guiding a family through hunger and disease. It doesn't matter whether the cultural hero-system is frankly magical, religious, and primitive or secular, scientific, and civilized. It is still a mythical hero-system in which people serve in order to earn a feeling of primary value, of cosmic specialness, of ultimate usefulness to creation, of unshakable meaning.**

There is a considerable amount of observational truth in this one paragraph, not the least of which is the assertion that cultural hero-systems are the same the world over, whether they are religious or secular in nature. All of these systems give us a "feeling of primary value," which goes to the essence of man's dilemma: his need to find value for himself even though he is an insignificant lump of corruptible flesh in an uncaring universe. It is especially noteworthy that Becker saw no difference between religious or secular hero systems, which is exactly what Joseph Campbell discovered in his research on human mythologies. The most shocking aspect of *The Hero of a Thousand Faces*, which affects people even today nearly seventy years after its publication, was Campbell's nonchalant inclusion of Jesus Christ as an ordinary figure of myth in the pattern of Osiris for the ancient Egyptians, or Dionysus for the Greeks.

Like the great thinkers before him in the field of psychology, and like Ernest Becker, Joseph Campbell had the ability to "think outside the box," to use a modern popular phrase, which meant that he could place himself outside of his inherited religious beliefs, or even his own natural intrinsic biases to believe in a God, in order to look at God-belief and religion skeptically and objectively. This made it possible for him to shed all personal reverence for the figure of Jesus Christ as a potential Lord and Savior, and instead see him through the lens of a sociologist, or psychologist. Jesus Christ could then be slotted into a mythological category, based on the mythologies which quickly sprung up about him after his crucifixion (Campbell was making no claim that Jesus Christ never existed, which is the core premise of today's Mythicist movement).

The benefit to Joseph Campbell of placing Jesus Christ into the context of a mythical hero project was that Campbell got the scholarship right. The cost was that *The Hero of a Thousand Faces* became, and remains today, a book approached with deep anxiety by Christians, many of whom have condemned it over the decades as irreverent and dangerous. Were Campbell alive today, he would have the satisfaction of seeing what has happened to modern American Christianity, and how it has transformed itself more into a secular hero project. Some of this is evident in the popularity of what is called the Prosperity Gospel, of which we will have more to say later in this book, and which postulates that if a Christian "sows a seed" by investing dollars in a pastor and ministry, God will multiply this investment many times in the here and now. God, as defined by the Prosperity Gospel, is the financier for each believer's hero project.

A more obvious case is the popularity of modern Christian preachers who are nothing more than secular motivational speakers, such as televangelists. Their use of the gospels is minimal; their Biblical quotes are highly selective and devoted to supporting a message of hope, encouragement, serenity, and placing one's faith in God. There is very little room in this sort of preaching for a traditional message of Christian charity; sermons are not devoted to how believers must spend some of their time helping others. These modern preachers, many of whom are either megachurch celebrity leaders who minister to congregations of 10,000 or more, or some of whom are televangelists, draw all attention to the personal, daily struggles of their followers. Sermons are devoted to managing personal relationships, overcoming depression, fighting off doubts, eliminating negative thinking, coping with setbacks, and keeping one's focus on moving forward rather than staying mired in the present or trapped by past failures. These are motivational lectures that any talented self-help guru could deliver, the difference being that the listener is sitting in what appears to be a church, and the lecturer is constantly emphasizing how each individual has a hero project that is ultimately in the hands of God.

We see here as well the process of transference at work. Celebrity megachurch pastors and televangelists, who have enormous followings, are masters at persuading their flock to transfer all of their anxieties from themselves to God. The way this is to done is to repeat over and over pithy, easily memorable sayings, and here are some common examples:

> "God has a bigger plan for me than I have for myself."

> "God's plans for my future are far greater than my fears."

> "God's plan is always the best. Sometimes the process is painful and hard. But don't forget that when God is silent, He is doing something for you."

"If your plans don't work out, don't worry. God has better ones for you."

"Faith is trusting God even when you don't even understand his plan."

"Dear God, if today I lose my hope, please remind me that your plans are better than my dreams."

Embedded in these aphorisms are the principles of personal and social meaning that were described by Becker, Campbell and so many other observers. These include not only the principle of transference, wherein the believer is encouraged to put faith in God, and allow all personal anxieties to be given to this mental image, but also the workings of the symbolic action system that Ernest Becker wrote about. In this system, religion is just one of many participants, since religion encourages the believer to understand that God (in other words, society) has plans that are "better than my dreams." If society considers that your gifts to civilization are those of a certified public accountant, whose services are quite valuable to the management of complex business operations, then religion encourages you to accept this as your fate in life, because you are also serving God in a way. When Jesus said "render unto Caesar that which is Caesar's," he was referring to more than taxation. He was establishing that each person has a responsibility to participate in society in some productive way.

The practices of dealing with death which we have discussed in this and the previous chapter – repression, Infantile Amnesia, the use of transference, and the establishment of our personal hero project – begin at birth and carry throughout the rest of our lives as adults. With this review, we have now completed our scientific study of how God-belief operates from the infantile to the adult stage of our lives. In the next chapter, which finishes Part One of this book, we will put together all of these pieces, to find the similarities and smooth out the overlapping parts, so that we can have a workable, comprehensive model of the origins of God-belief and religion.

CHAPTER EIGHT

The Glass Composite Model
for God-Belief

We are God's chosen people.
We are God's treasured possession.
Let us rise in mighty strength to possess our rightful
places as God's children.

— Lailah Gifty Akita

The Five Stages of God-Belief

The author quoted above, Lailah Gifty Akita, is a Nigerian Christian who writes inspirational books for Christian believers. This is a very common genre in modern Christianity, suitable for social media with its emphasis on memes and tweets, and popularized by televangelists and megachurch celebrities who are little different from secular motivational speakers. The short, and easy-to-remember bromide – of the sort we just discussed in the previous chapter - is an important tool for motivational speakers, but these aphorisms sometimes are perfect examples of the contradictions inherent in God-belief and religion.

Ms. Akita urges her readers to "rise in mighty strength" so that they can assume their rightful place "as God's children." The image of God's children having mighty strength, as soldiers for God, is both contradictory and silly. If there were such a thing as God's Army on earth, it would be composed of adults doing battle against evil. But what we have learned in Part One of this book is that God-belief arises from and derives its potency from our infantile and childhood experiences. As such, a believer in God will always be a child to God the Father, or like an infant to Allah, or progressing through multiple reincarnations in Hinduism in order to reach the divine maturity represented by Nirvana.

It is with this understanding that we can now summarize all that we have learned about the origins of God-belief in our infant and childhood years. We can create a composite model from all the scientific work that we have reviewed, and in this model there are five progressive stages we all experience which lead us to God-belief as the default condition for all adults. For lack of any other specific way to describe this model, we are going to call this the Glass Composite Model for God-Belief. This will be my somewhat self-serving effort to distinguish this model from all the others we have reviewed so far; consider it my personal hero project for helping the reader make sense of all the marvelous research that has cropped up in recent years about God-belief. We start with where God-belief first occurs, at our moment of birth.

Stage One: The Neural Imprint of Template God

Extreme helplessness breeds acute dependency. We are born into this world unable to feed ourselves, stay warm, keep ourselves clean, or even move about in order to avoid predators. We have to be picked up and carried from place to place, and while these conditions are similar to those experienced by newborns of many species on this planet, we humans are unique in the extraordinary length of time it takes for our offspring to become independent.

The *only way* we as infants survive at all is through the constant care and attention of adults. The reason we as a species require such a lengthy period of development is the complexity of our brain. Other vertebrates, from our immediate cousins the Great Apes down to the lowly lizard, do not need a brain anywhere near our size or complexity, because they do not have language skills, and they do not experience the rational thought or range of emotions that characterize our species. Our offspring take nearly a year to learn to walk, another year after that to begin to learn language, and another two or three years after that to obtain some degree of self-awareness. Complete development, so that our children can truly survive on their own in a post-industrial society, requires another fifteen or more years of education.

Over hundreds of thousands, if not millions of years, we humans have developed a survival instinct that is activated at the moment of our birth, and that instinct is very simple: we must attach ourselves as an infant to a caregiver who will nurture and protect us in our helpless state. As infants, we are provided with powerful motivators that urge us to repeat the experiences necessary for our survival. Our brain issues a pleasurable chemical reward to our neurological system whenever we feed at our mother's breast, and whenever we are held tightly by our caregiver and protected from harm. We *feel* the love of our caregiver, as the areas of our brain which govern a sense of security and which generate feelings of calm are activated when our caregiver does what she is designed to do – care for us.

But just as important to this process are the changes which affect adults in the presence of an infant. Adults are hard-wired to find the cries of an infant as irritating – as something that needs to be stopped, and stopped now. Adults are motivated to pick infants up to stop them from crying, which benefits both the infant and the adult. Adults are also fascinated by infant faces, and they are primed to want to interact with the infant, to welcome the infant to the human species and to help it participate in the broader world.

But there is one adult in particular who is of the greatest importance to us in our infant state, and that is our primary caregiver, the person who is prepared to devote almost all their waking time to our needs in our early years as an infant. That person is for most of us our mother. It is to this individual we as an infant most strongly bind, or connect, at the moment of our birth. Numerous scientific studies have proven the extreme importance of these first few minutes after birth, when we as a newborn are placed in the arms of our mother. The flood of comforting emotions which overwhelm the infant at that moment, are matched by the same emotions which overwhelm the mother, and which instruct her to say to herself: "taking care of this infant as completely as possible is my job now, and this is what I want to do."

This is the process of creating a neural imprint between the mother and child. The adult in this process need not be the mother. Until the past century or so, childbirth was a highly risky business, and many mothers died as a result of childbirth. Our species coped with this fact by allowing the infant to bond with another adult, and to allow most adults the potential to bond with an infant.

Thinking about this from the infant's perspective, as a newborn we will have one-quarter of the brain at birth that we will have as an adult. This "lizard brain" is our inheritance at birth, and it consists mostly of the Brain Stem and the Limbic System. As infants, we have enough of a brain to eat, digest food, know where our food comes from (the mother's breast), and know how to scream loud enough to get the attention of adults when we need something. This is the brain necessary to survive as an infant, but almost immediately, our two other key components of our brain develop as well. Our Cerebellum allows us to walk upright, and our Cerebral Cortex begins to give us the mental capacity necessary for us to survive as an adult. But since it takes years for the Cerebral Cortex to fully develop, our progress from infant to child to adult takes years as well, and this is a journey from complete helplessness to lesser degrees of helplessness.

In this journey, helplessness is always accompanied by dependency. At our infant stage, our extreme helplessness creates acute dependency on our caregiver, and because of this extreme helplessness, the characteristics we perceive in our caregiver are extreme as well. Our caregiver, meaning most often our mother, is all-powerful, all-knowing, always present, and all-loving. She is our Creator and the obvious creator of the world around us. She is also, by definition, always right. As infants, if we perceive our caregiver to be unhappy with us, we interpret this situation as our fault, not hers, because she is all-powerful.

It has to be emphasized that this process of bonding with a caregiver imprints in our own mind, likely somewhere in our Mid-Brain or Limbic System (which are the major functioning parts of our brain at birth) the fact that *there exists in this world a real individual of flesh and blood, whom we can touch and with whom we can interact, who is all-powerful, all-knowing, always present, all-loving, and our Creator.* Exactly how this process takes place is open to conjecture, but it may be very similar to the neural imprinting that takes place between a newborn and its mother.

Research has shown that a large surge of oxytocin, a natural hormone which acts to reduce stress-inducing steroids such as cortisol, is released in the newborn's brain the first time it breastfeeds. Additional emissions of oxytocin are activated with continual feeding, and as the infant is touched, caressed, or held closely by its mother and other caregivers.

Breastfeeding also stimulates the production in the infant of prolactin, a protein which contributes to a sense of well-being and trust. Norepinephrine, otherwise known as adrenalin, is secreted through breastfeeding and is activated in the brain through regular touch from caregivers, providing the infant with increased attentiveness to its mother and its environment. Additional research indicates an infant's brain is reorganized to manage stress better, the more the infant receives positive hormonal and protein infusions through constant attention by adults, especially through breastfeeding, touch, comfort, and audiovisual stimulation.

In this sense, the infant's brain in its first year or so of life, becomes "hard-wired," not to believe that there is a God, but to be conditioned with the impression that there exists an all-powerful, all-knowing, always present, all-loving Creator who can be summoned to come to the aid of the infant when called. This impression is a template for God's most fundamental qualities, and the template serves as a foundation and basis from which the infant, having many years later reached adulthood, can put its faith into an invisible superhuman who is called God. We therefore will call this impression, born of thousands of positive and rewarding experiences with a mother and other caregivers, **Template God**.

Freud would say this template lies in our unconscious mind, because it operates "behind the scenes." The template is our impulse for belief in God, and the fundamental source of God-belief. It is the one aspect of God-belief which we share with all other humans, because as an infant, the template is extremely general, providing us not so much a visual image of God, as an emotional impression of someone we later will call God, to whom we can always turn. The template is also the reason we can assert that belief in God is the default position for humans. This generalized impression of a God-like figure exists in all of our minds, simply because we survived childhood. Template God, as our most basic impression of a God-like figure, is the one model for God that all humans share, because Template God is survival God as well.[2]

We can begin charting the main aspects of our model for God-Belief, starting with Template God in Chart Nine below. In this chart, the time line at the bottom of the chart, ranging from birth to adulthood, shows where the stage of development begins. In the case of Template God, the process begins at birth. Notice that on this timeline, a critical development point occurs around age two, when language begins to be explored by the infant. Language ability is considered by many child development experts to be the dividing line between infancy before language, and childhood after.

---

2   Carl Jung, in his work on the collective unconscious, developed a model for a universal archetype, or template, operating at the lowest level of the unconscious. "A more or less superficial layer of the unconscious is undoubtedly personal. I call it the 'personal unconscious.' But this personal layer rests upon a deeper layer, which does not derive from personal experience and is not a personal acquisition but is inborn. This deeper layer I call the 'collective unconscious.' I have chosen the term 'collective' because this part of the unconscious is not individual but universal; in contrast to the personal psyche, it has contents and modes of behaviour that are more or less the same everywhere and in all individuals." Carl Jung. *The Archetypes and the Collective Unconscious*, 1934, Princeton University Press. (Jung, 1934)

Notice three other points about these charts. Most of the stages of development have a beginning point that is applicable to the average child. Many children, for example, will enter into the second, third, fourth or fifth stages many months before or after the chart indicates occurs on average. Except for the fourth stage, all of these developmental stages leave impressions which carry on throughout our adulthood, and affect us at an unconscious level until we die.

Second, at the very bottom is a spectrum of perception of God which begins in infancy at the "reality" stage of perception. In other words, Template God is based on a being or beings who are very real to us as infants. Gradually, however, all these developmental stages fade away from our conscious mind and contribute to a pre-adult and adult perception of God as an illusion. Like any illusion, God can seem real to us only to the extent we wish ourselves to accept him and believe in him as real.

Third, when we say that Template God, or any of the other versions of God that will follow next in the model, is "Located in" the Mid-Brain, Limbic System or any other area of the brain, we are conjecturing a possible source of the impressions infants and children create of these God-like figures. There simply isn't enough research on these questions to say definitely where these impressions are firmly situated and how these elements of the brain process, store, and exchange such information with other areas of the brain. We are making reasonable estimates based on the opinions of researchers, or the current understanding of how the components of the human brain function.

Chart Nine: The Neural Imprint of Template God

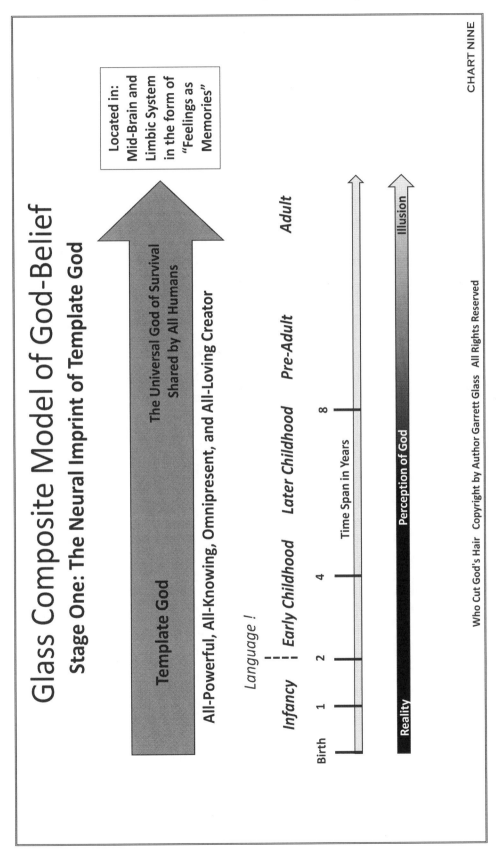

Glass Composite Model of God-Belief

Stage One: The Neural Imprint of Template God

Template God

All-Powerful, All-Knowing, Omnipresent, and All-Loving Creator

The Universal God of Survival Shared by All Humans

Located in: Mid-Brain and Limbic System in the form of "Feelings as Memories"

Language !

Infancy | Early Childhood | Later Childhood | Pre-Adult | Adult

Birth 1 2 4 8

Time Span in Years

Perception of God

Reality

Illusion

CHART NINE

Who Cut God's Hair   Copyright by Author Garrett Glass   All Rights Reserved

Stage Two: The Discovery of "They" as the Basis for Anthropic God

Within six months of our birth, our brain has changed radically. The components of the Limbic System, for example, take on greater responsibilities, just as the Cerebral Cortex – the gray matter on the surface of our brain - is beginning to develop at a very quick pace. As part of this development, the synapses which connect the neurons of our brain with each other are also emerging. It is now possible for the different parts of our brain to communicate with each other. The image of our caregiver as perceived by our Occipital Lobe can now begin to be stored and processed in the Hippocampus as a memory, and the image can also be interpreted by our Parietal Lobe. The Parietal Lobe is not the only area of the brain which processes such images, but it has a principal role in the cognitive interpretation of the images we endow with "otherness," defining these images as entities distinct from ourselves.

Since our Parietal Lobe provides us with a sense of space, once the Parietal Lobe begins to develop, we can perceive that there is an external world, apart from our body. We are going to need this facility when we begin to walk, which is a process that commences in fits and starts, by standing up, holding on to objects as we negotiate our feet from place to place, and eventually letting go of these objects as we let our feet propel us forward. We are still wobbly, and we will bump into things from time to time, and we will fall down and hurt ourselves, but absent any developmental impediments, we all learn to walk just fine.

The same can be said regarding our sense of identity. At birth, we do not perceive ourselves as "I" – we simply exist in a state of extreme helplessness. With the benefit of a maturing Parietal Lobe, we begin to see that the faces of humans around us represent individuals who are physically and emotionally distinct from ourselves. We learn that we are distinct from them: we are an individual. We have an identity all our own. We are "I," and there is no other "I" but us. Everyone else is "They," though over time, each of these individuals will be perceived as distinct and unique.

Of all those who constitute "They," our mother, or our primary caregiver in her absence, is still the most important figure in our lives. She is the source of our template for God – of Template God. Since she is now beginning to appear to us as a distinct and unique individual, our perception of God becomes invested in her, and takes on her characteristics. This process does not alter Template God. The template is a very general construction of a God-like figure who is always available for us with all of her infinite powers. Template God is not killed off or altered in any way as we mature beyond infancy. Instead, it is augmented with a *composite of impressions* which consist of visual images, auditory and physical experiences, smells, and memories of our caregiver, all of them physical characteristics. We will call this our **Anthropic God** – the God who is now perceived by us as a human entity, and one who is distinct from us physically, in an external reality different from ours. *Anthropic* refers to that which defines and shapes the universe through human experience, or which allows us to understand the universe through the human mind.

There are two critical aspects of this process which play an important role in our adult perceptions of God. The first is the association of God with our mother, and to a degree, with the other adults who interact with us, like our father. All of these people are God-like figures, which informs us of a very simple fact: *God has human characteristics.* This evolves into the concept that God is one of us, a member of our species, but an all-powerful version of a human. This is the process of anthropomorphizing God, which means giving him human characteristics. It will play a role in all future interpretations of God for our species. The gods of the forest or the seas had human characteristics for our primitive ancestors, just as the Egyptian gods which were half animal were also, in terms of their character, fully human. To have any true importance for us as adults, God must be anthropomorphized, or we cannot relate to him.

The second important aspect of this process of infant development, stemming again from the capabilities vested in our Parietal Lobe, is that our caregivers are now perceived by us as physically distinct from us. Not only do we have an "I" to be countered to their personal identity, we have a body that is different from their body. As these humans are part of the external world, and as they are representatives of our Anthropic God, so too does Anthropic God become part of the external world. *We begin to think of these God-like figures as "out there," in the real world, as externalized figures.* As we mature, we will come to interpret the real world in much grander terms, as the cosmos. God becomes associated eventually with the cosmos, or universe, because only an all-powerful figure can reside in something as awesome as the cosmos.

Chart Ten: The Discovery of "They" as the Basis for Anthropic God

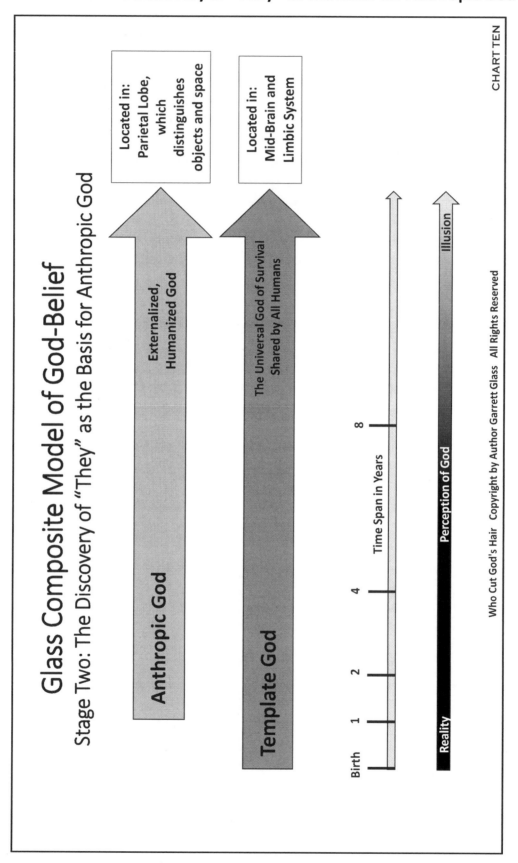

Glass Composite Model of God-Belief

Stage Two: The Discovery of "They" as the Basis for Anthropic God

Located in:
Parietal Lobe, which distinguishes objects and space

Located in:
Mid-Brain and Limbic System

Anthropic God

Externalized, Humanized God

Template God

The Universal God of Survival Shared by All Humans

Illusion

Birth   1   2   4   8

Time Span in Years

Reality

Perception of God

CHART TEN

Who Cut God's Hair   Copyright by Author Garrett Glass   All Rights Reserved

Stage Three: The Good and Bad Versions of Behavioral God

As we approach childhood, beginning around age two when most infants develop language skills, we have already stored up numerous impressions of Anthropic God. This God is much more complex than Template God, who is a construct of "feelings as memories." Anthropic God takes on different human physical characteristics, whereas Template God is a formulation common to all humans. At some point, our Anthropic God begins to differ from some other child's Anthropic God, because the individuals who help create Anthropic God are physically different from the parents or caregivers of any other child.

As Stage Three develops, we have entered early childhood, generally defined as age two and beyond for several years as we learn language. We are more perceptive of the behaviors and emotional makeup of our parents or caregivers. This new knowledge helps us create a more refined and more human God with good and bad behaviors. We refer to this God as **Behavioral God**, because he begins to reflect distinctive human behaviors that evoke emotional reactions within us.

Earlier in our discussion of the research conducted by Melanie Klein and her followers, we referred to our infant self as taking on the fault of any imperfections which may be perceived in our caregiver. The infant creates a paired relationship with a Bad Mother whose behavior is hurtful occasionally, but tolerable overall. For example, if our mother is late in picking us up when we are hungry or uncomfortable, this cannot be perceived as the fault of our mother. She is all-perfect, being all-powerful. Her love is perceived by us as boundless. Psychoanalysts describe this process as one of self-blame; the infant takes on the faults of the mother, which will inevitably occur, and accepts blame for these faults, because the contradiction between an all-perfect yet flawed caregiver is too great to comprehend or bear.

This process plays out as infants approach childhood, which brings with it a capability to understand to some degree adult talk and adult actions, as well as a growing perception by the child of their own identity as distinct from any others. In childhood, the flaws of the caregiver, or the caregivers for that matter, are much more obvious and cannot be avoided. The mother and father might argue on a regular basis, and the emotions generated in these arguments are ugly and to be feared. For some children, the presence of a newborn in the family may be very traumatic, because parental attention is now drawn more to the newborn than to the child. Behavioral God is perceived as having favorites, and the question we as a child must ask is: whose fault is that?

It cannot be the fault of Behavioral God – that is a contradiction in terms. It must be our fault, as the child. We did something wrong to lose the affection of our caregiver, or to cause our mother and father to be angry with each other. It does seem rather odd that a child would blame themselves for the problems of the parents, but nearly one hundred years of psychotherapy by countless psychiatrists, psychologists and psychoanalysts has borne out the observation that adults can develop significant psychological problems which stem from childhood causes such as sibling rivalry, lack of affection from a parent, or emotional strife in the family.

The phenomenon at play here is psychological splitting, the consequence of which is *childhood guilt*. Guilt leads to self-blame, and self-blame to self-worthlessness. Nowhere is this feeling of self-worthlessness more evident than in the relationship of the child to its Behavioral God. The pattern that is established is a life-long relationship of an unworthy child to its perfect Behavioral God. Just as we in our actual lives will always be and act like children in the presence of our parents, so to the extent we have a religious or spiritual life we will always be and act like children in the presence of that figure we call God. We cannot but be children in the presence of God, because we are flawed, and he is perfect.

Putting this into the context of our model for Behavioral God, we find that as children we have at least two versions of this God, the Good God and the Bad God, both of them sub-sets of Behavioral God. The Bad God is the paired relationship we create that represents our interactions with the parent or caregiver on whom we are most dependent, and whose bad behavior, while occasionally hurtful, is tolerable.

If we have a remote father, or a father who physically or emotionally abuses us, those experiences form the basis for the Cruel God, yet another subset of Behavioral God. We can see how this has affected the Judeo-Christian God. The God of the Old Testament has a number of characteristics of a Cruel God – jealous, vindictive, prone to anger, and vio-lent. The model for the Old Testament God might have been the Cruel God of the writer of Exodus, for example, or more than likely, the composite of a culture which prompted its children to create similar models of Cruel Gods from their own experiences with their caregivers.

To make a rather speculative historical observation, it might have been the case that the Hebrews of circa 1400 – 1200 BCE, a nomadic group militarily beset at all sides by the Assyrians, Hittites, Canaanites, and Babylonians, did not have the possibility of treating their children with regular patience and attention. The culture might well have promoted in children a model of a Behavioral God who was aloof and valued defensive warfare and all its cruelties. By the Roman Era, there were communities of prosperous Jews and other cultures living in urban settings throughout the Roman Empire. Parents under these circumstances had the luxury of devoting time and love to their infants and children. The culture of the Roman Empire was ripe for a different model of Behavioral God – someone who was comforting, protective, a provider of all of one's needs, patient, yet demanding of respect. The type of God Jesus of Nazareth was describing – "Abba," the Father – fit this definition very well, and was a revolutionary change from the Hebrew God of the Old Testament. No wonder the message of Jesus Christ spread quickly throughout the Roman Empire: millions of urban dwellers could sympathize and comprehend this type of God.

In the first of the following charts, we add Behavioral God to our model for the development of God-Belief. The arrow in this chart shows that the development of Behavioral God begins at age two. This is an assessment of the average experience of a child; some children can begin to interpret the good and bad behavior of their parents or caregiver earlier, depending on when they develop language skills.

In the chart immediately after, we show one example of the psychological "splitting" process which can occur at this time: the creation of a Good God whose behaviors are all-benevolent, and the creation of a Cruel God whose behaviors are harmful. We could substitute a Bad God relationship for the Cruel God, and the chart would look the same. We've kept this chart simple and used only three examples of paired relationships, but for any particular child, there could be other such relationships which bring about psychological splitting. At the bottom right of this chart is a box representing the residuum of this process, which is what is left of our own personality. What is left is the psychological dross, which is the morally deficient, spiritually weak individual we feel we are meant to be – the sinner who is estranged from our own creation of the idealized, perfect human being that is God. This is the guilt we take on our own shoulders, at a very early age, for the bad but tolerable behavior of our Bad Good, or worse still, for the cruel and intolerable behavior of our Cruel God, if we are so unfortunate as to have such an experience.

We must note that as Stage Three progresses, we have moved from infancy into early childhood, and we now maintain three basic impressions of God: Template God, Anthropic God, and Behavioral God. We call these impressions "God," but this is only a label, and this is not to say that at age two, three or beyond, a young child will look at their parents or caregivers and say, "Aha! You are God!" A child at this age hasn't the conceptual ability to understand the meaning of the word "God," other than whatever it may hear from its parents or Sunday school teachers about a third party that exists outside of its family structure. What the child does have, through its experience with its parents and caregivers, is a strong sense that there exists in this world real beings who are all-powerful, all-knowing, always present, and all-loving. These beings appear to have created the world and everything in it, including of course the child, and the child senses or appreciates that it is utterly dependent on these beings for its continuing existence. Gradually these beings assume human characteristics and an identity separate from the child, and which places them "out there" in the external world. As Stage Three progresses, these individuals are found to have human behavioral and emotional characteristics, many of which are loving and benevolent, but some of which appear to the child to be hurtful, capricious, or even malicious. The charts below show us what occurs at this point, and how young children assume personal guilt, as a means of reconciling the "bad" behaviors of these God-like individuals with the omnipotence and perfection the child had earlier perceived was their nature.

Chart Eleven: The Good and Bad Versions of Behavioral God

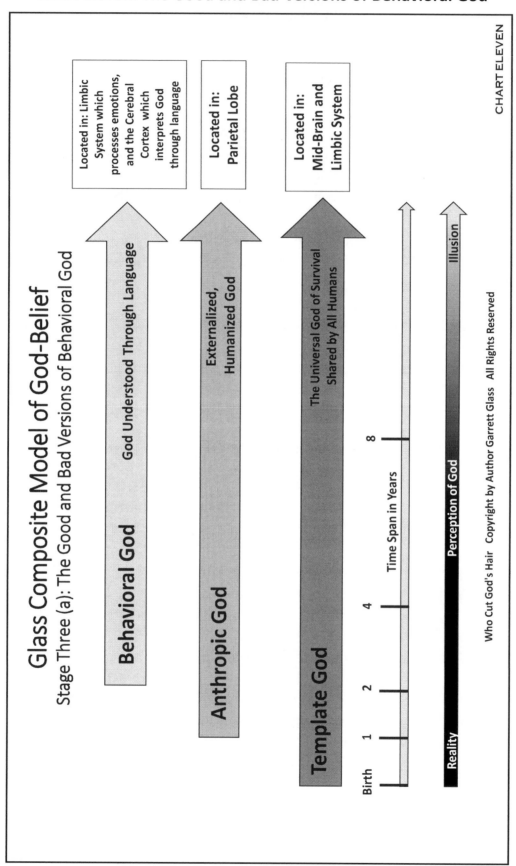

Glass Composite Model of God-Belief
Stage Three (a): The Good and Bad Versions of Behavioral God

Behavioral God — God Understood Through Language

Located in: Limbic System which processes emotions, and the Cerebral Cortex which interprets God through language

Anthropic God — Externalized, Humanized God

Located in: Parietal Lobe

Template God — The Universal God of Survival Shared by All Humans

Located in: Mid-Brain and Limbic System

Birth 1 2 4 8

Time Span in Years

Reality

Illusion

Perception of God

CHART ELEVEN

Who Cut God's Hair   Copyright by Author Garrett Glass   All Rights Reserved

— 98 —

Chart Twelve: The "Splitting" of Good and Bad Behaviors

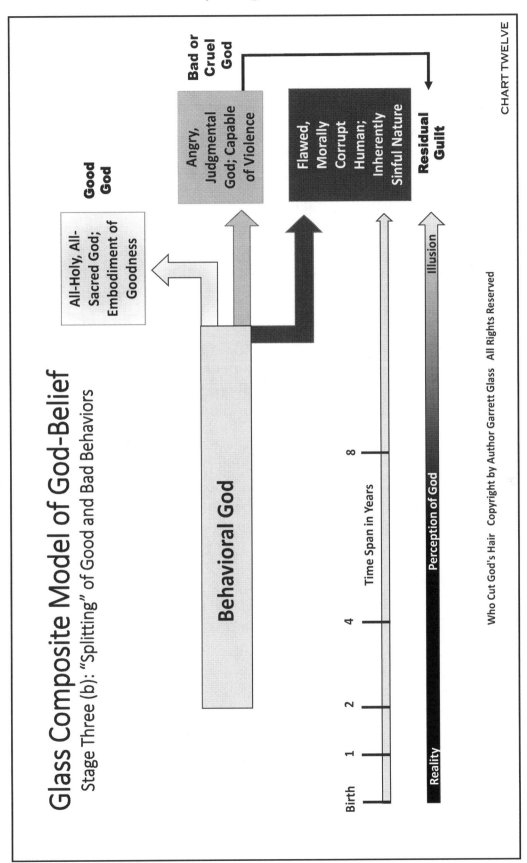

Glass Composite Model of God-Belief
Stage Three (b): "Splitting" of Good and Bad Behaviors

Good God

All-Holy, All-Sacred God; Embodiment of Goodness

Bad or Cruel God

Angry, Judgmental God; Capable of Violence

Flawed, Morally Corrupt Human; Inherently Sinful Nature

Residual Guilt

Behavioral God

Illusion

Perception of God

Reality

Time Span in Years

Birth   1   2   4   8

CHART TWELVE

Stage Four: The Unmasked God

At some time starting around age three, but more than likely around age five, we as children are now cognizant of the real world, and we are more endowed with adult-like Frontal Lobes. We can conceptualize and understand the fundamental truth of human existence: that it is limited, or finite. We come to understand what death is, that it will occur to our parents, and then to ourselves. This revelation is, as Ernest Becker reminds us, "the terror" that plagues our mortal being.

Along the way to our understanding of the finality of death, we have already begun to comprehend the limitations of our parents, or caregivers. We are witness to their flaws, and occasional victim to their capricious moods. We internalize all this as children, taking the blame for their behavior on to ourselves, and suffering the guilt of our worthlessness as a result. We psychologically protect our parents and ourselves by creating a bad paired relationship which we then project into the external world and pretend that it happened to someone else, perpetrated by some other parent. The guilt associated with this relationship, nevertheless stays with us.

Up to this point, our parents are still God-like figures in our eyes. They are the model for our Behavioral God, and we take all of our experiences with them and compartmentalize them into a Good God and a Bad God. Both of these conceptions are still imbued with the powers of God. Even the Bad God is still a Creator figure, with omnipotence, omniscience, and omnipresence.

All this is to change drastically when we learn about death and the ineluctable death of our parents. With this knowledge, we view them no longer as gods, but as limited creatures like ourselves, fully human with all the weaknesses that implies. Suddenly, our model of Behavioral God, both the good and bad versions, evaporates before our eyes, as does the trust we placed in our parents or caregivers as God-like figures.

Which is the more painful realization: that our parents and we ourselves are condemned to death by the mere circumstance of the gift of life; or that our parents are no longer God-like figures, and that the real-life experiences we had with them have been illusions, because we had invested qualities of perfection upon people who ultimately proved to be as flawed as we are? For those social scientists who work in the field of Terror Management Theory, the answer is obvious. As children, we are deeply traumatized by our confrontation with the truth regarding our mortality, and nothing can be more frightening than this discovery.

But some thought must be given to the trauma of iconoclasm which we also suffer when we learn about death. With that discovery we learn as well that our parents have feet of clay, metaphorically and literally, in that ultimately they shall return to dust. At the same time, we have been deceived about them, or we have deceived ourselves into thinking they are gods, neither of which is a comforting thought. And what does all this mean for our child-like understanding of God? The one thing we know for certain is that God – meaning the images we have created that constitute an entity we will ultimately call God - does not really exist in the form of our parents or caregivers. Where is he? Does he exist at all?

All this confusion, which plagues every child once they learn about death and the all-too-human limitations of their caregivers, is ultimately swept beyond the reach of our conscious mind by the coping mechanisms we humans have devised to deal with such massive psychological trauma. In our discussion about death in the previous chapters, we looked at the biological phenomenon of Infantile Amnesia, whereby our brain doesn't erase our memories of these painful revelations, but rather deactivates the synapses which allow us to bring these memories to the surface, to our conscious mind. Along with our anxieties surrounding death, our brain engulfs into Infantile Amnesia our ability to recall our memories of Template God, Anthropic God, or Behavioral God (Good God, Bad God, or Cruel God). The memories are still there, lodged this time not simply in our primitive brain, but in our more mature brain, including the Hippocampus, as well as parts of the Cerebral Cortex that interact with our Limbic System.

Template God is more than likely lodged in our primitive brain, consisting of the Mid-Brain. Template God still works upon us from our unconscious mind, in a remote way, as an impulse which informs us from "behind the scenes" that there is validity to the concept of an all-powerful, all-knowing, always present, all-loving, and Creator figure. Anthropic God is still there as well, operating from our unconscious mind, providing some rational justification to the concept that God in human form must exist as an entity separate from ourselves, out in the cosmos, dwelling in a place where death does not lurk, and where all our cares and anxieties are vanquished.

The absence of Behavioral God in our childhood lives is a terrible blow. Our parents go off to complete their mortal existence, having lost their divinity in our eyes, and we have lost any connection to a real, God-like figure. Throughout this process, all three of the Gods of our infancy and childhood have been unmasked, and shown to us as they really are: all-too-human, flawed, and utterly mortal beings who cannot possibly serve as the idealized human being we have been constructing as God

We call this process the discovery of **Unmasked God.** He is the last God in our life who will be truly real, because he is such a frightening disappointment. He is a god only in the sense that our parents or caregivers still have enormous power over us and our dependency on them for our survival is undeniable. Yet he is not a god because no god would ultimately desert us, leaving us alone to fend in a dangerous world. Unmasked God leaves a terrible void in our psychological and emotional make-up, and he is the source of what Christians call "the God-shaped hole in our heart." We are not forced or compelled to fill this hole, but it is a very intense longing which most humans satisfy by creating a new God out of all the previous manifestations shown to us in the earlier stages of development. This process is pushed along considerably, by the way, when our parents or caregivers step in and introduce us to this new God, through religion. We will explore the nature of this new God, who if we choose will serve our needs for the rest of our life, in Part Two of this book.

The next two charts update our model, and then show the effect of Infantile Amnesia, Repression, and Transference, as they force our conscious knowledge of Template God, Anthropic God, and Behavioral God to fade away to nothing more than impulses which solidify God-belief as the default position of mankind in the face of the terrible knowledge of our mortality and that of our parents or caregivers.

Chart Thirteen: The Unmasked God

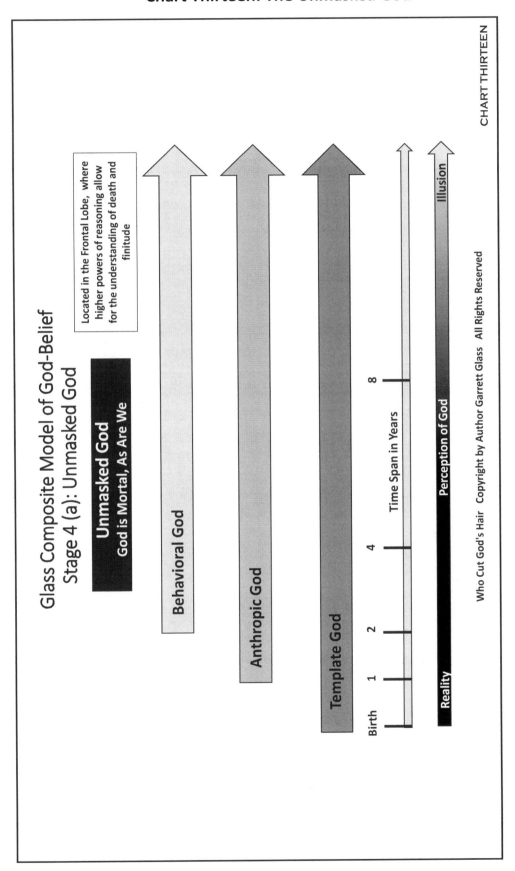

Glass Composite Model of God-Belief
Stage 4 (a): Unmasked God

Unmasked God
God is Mortal, As Are We

Located in the Frontal Lobe, where higher powers of reasoning allow for the understanding of death and finitude

Behavioral God

Anthropic God

Template God

Birth 1 2 4 8

Time Span in Years

Illusion

Perception of God

Reality

CHART THIRTEEN

Chart Fourteen: Denial of Death

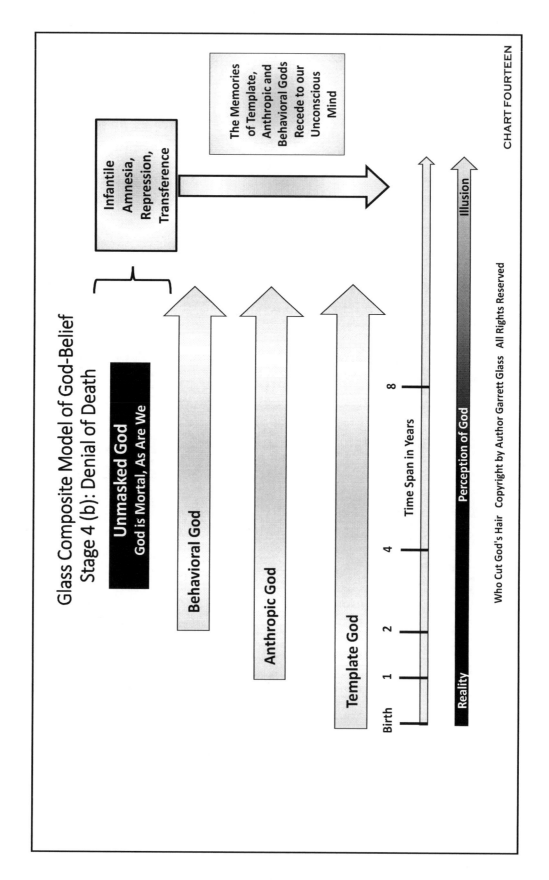

Glass Composite Model of God-Belief
Stage 4 (b): Denial of Death

Infantile Amnesia, Repression, Transference

The Memories of Template, Anthropic and Behavioral Gods Recede to our Unconscious Mind

Unmasked God
God is Mortal, As Are We

Behavioral God

Anthropic God

Template God

Illusion

Birth 1 2 4 8

Time Span in Years

Reality

Perception of God

CHART FOURTEEN

Who Cut God's Hair   Copyright by Author Garrett Glass   All Rights Reserved

Stage Five: The Path to Not-Death through Illusory God

Unmasked God creates a longing for a replacement God, a figure who can still provide all the solace and protection of the three previous embodiments of God we have come across in our young life, despite the fact that we now lack any evidence that this replacement God could ever exist in the real world we inhabit. We are thus primed to create yet another manifestation of God – **Illusory God**. As the name implies, Illusory God is an illusion, meaning he does not exist in the material world. Or, like a quantum particle in physics which is said not to exist in reality until we humans decide to measure it, so God exists only at those moments when we humans contemplate him with our conscious mind. And like all illusions, as Freud taught us, Illusory God represents a wish we desire to be fulfilled. In this case, it is the wish that God does indeed exist in a material, anthropomorphized, and externalized form.

In the previous chapters, we learned that when we arrive at the conclusion that our life is finite, and that we shall someday die, we repress this knowledge so that we can go about our daily life with some degree of equanimity. But while we can repress the idea of death, we cannot avoid it, and so we enter into a lifelong struggle against death. It is not enough, in other words, to ignore death; we must conquer death. We must achieve not-death. The question is, how to accomplish not-death?

One obvious answer, unsurprisingly, is to embrace fully the idea of Illusory God, because Illusory God fulfills our wish for not-death. How does this occur? When we learn about death, we learn as well about finitude as distinct from the infinite. The knowledge of finitude provides us with an understanding of time, and these discoveries occur close together if not simultaneously. As infants we had very little if any understanding of time, and because of this our Template God was assumed to have immortality, just as we as infants assumed our existence was eternal. As our Parietal Lobe developed we began to understand something more of time – that one day was succeeded by another, and "tomorrow" meant something significant. Still, there appeared to be an infinite number of tomorrows ahead of us in our childhood, until such time as we connected the death of our parents to our own inevitable destruction. It was then that we learned that for us, ultimately there would someday be one tomorrow which would be followed by no other tomorrows.

Unmasked God brings into our life Time, and Time, as well as Nature, are now understood to be our mortal enemies. If we wish to cheat death, one obvious solution is to conjure up the God of our infancy and childhood, because where that God existed, so too existed the immortality we assumed was our divine right as infants. The importance of Illusory God is not just that he is a comforting, protective, all-wise and all-loving being; it is that Illusory God carries about his being and presence the immortality we long for as a means of achieving not-death.

Note too that this longing is pressed upon us because God was once a real figure in our life, and heaven was our real abode. The God of our infancy and early childhood was not a fiction, and neither was heaven. Should we be at all surprised that as we mature from early childhood to young adulthood, we are mentally prepared to accept the conscious, rational argument that God still exists, along with heaven, only now he is out of our immediate presence? He is very human, but he is also physically very remote, to the point that no one can actually see him. This is what many of the great religions teach – that God is invisible, a spirit of infinite power. Why do humans around the world, whatever their religion, believe that God cannot be seen? Because each human, individually, can no longer "see" God or heaven through our conscious memories of our experiences with him as an infant or in childhood. Those experiences are all invisible to our conscious mind, because we have repressed them in order to avoid confronting two painful realties. The first is the terror we feel whenever we directly face up to our ultimate death. The second is the trauma we experienced when we realized that our parents, or caregivers, were not gods after all, and that we had been greatly deceived on that point.

If our parents or caregivers can no longer be seen as gods, and yet if we desire for God to still exist and provide us both eternal life and the removal of all our earthly anxieties, then Illusory God will have to do. In most cases, our parents or caregivers help this process along by educating us about Illusory God, only they don't call him that. They simply call him God, and state that he is very real, as is heaven and the afterlife. From age four to eight or a bit later, when our brain is working actively to repress all surface memories of our infant and early childhood years, it is more than just a convenience that our parents now intervene and tell us God and heaven both exist. Our parents have a serious motivation for doing so. They have long since come to accept their mortal limitations, and they worry about leaving their children alone in the world as orphans, especially if they themselves have lost one or more of their own parents to death and know the pain of true orphanhood. Illusory God becomes the ideal surrogate parent to pass along to their children, to watch over them and to protect them once both biological parents have passed on into death. Of course, what the parents pass along to their children is not viewed by them as an illusion at all, but as something very real, because that is what God feels like to them, and because that is what their parents or caregivers taught them.

This next chart summarizes the qualities invested in Illusory God, and the chart thereafter summarizes all five stages of the development of God-belief. If we accept the default belief in Illusory God, he will serve our needs, if we so choose, for the rest of our life.

Chart Fifteen: Illusory God: The Default Belief of Mankind

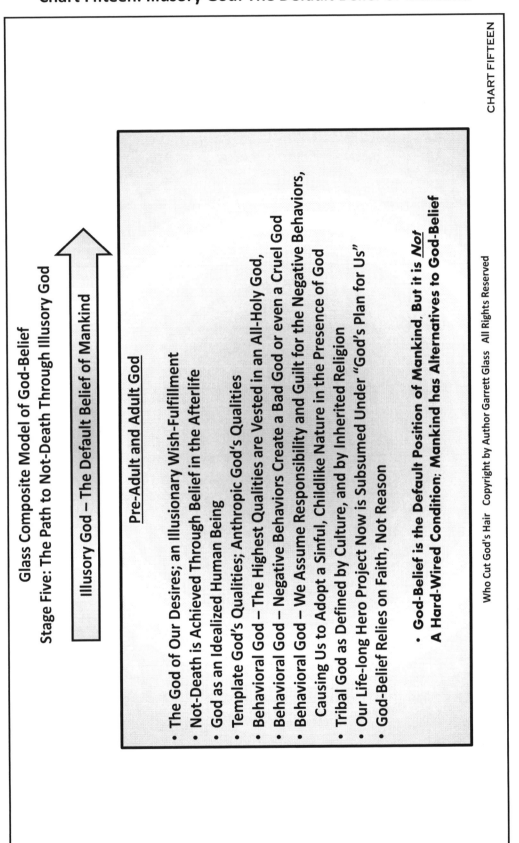

Glass Composite Model of God-Belief
Stage Five: The Path to Not-Death Through Illusory God

Illusory God – The Default Belief of Mankind

<u>Pre-Adult and Adult God</u>

- The God of Our Desires; an Illusionary Wish-Fulfillment
- Not-Death is Achieved Through Belief in the Afterlife
- God as an Idealized Human Being
- Template God's Qualities; Anthropic God's Qualities
- Behavioral God – The Highest Qualities are Vested in an All-Holy God,
- Behavioral God – Negative Behaviors Create a Bad God or even a Cruel God
- Behavioral God – We Assume Responsibility and Guilt for the Negative Behaviors,
   Causing Us to Adopt a Sinful, Childlike Nature in the Presence of God
- Tribal God as Defined by Culture, and by Inherited Religion
- Our Life-long Hero Project Now is Subsumed Under "God's Plan for Us"
- God-Belief Relies on Faith, Not Reason

- **God-Belief is the Default Position of Mankind, But it is _Not_
   A Hard-Wired Condition: Mankind has Alternatives to God-Belief**

CHART FIFTEEN

Chart Sixteen: Glass Composite Model of God-Belief – All Five Stages Represented

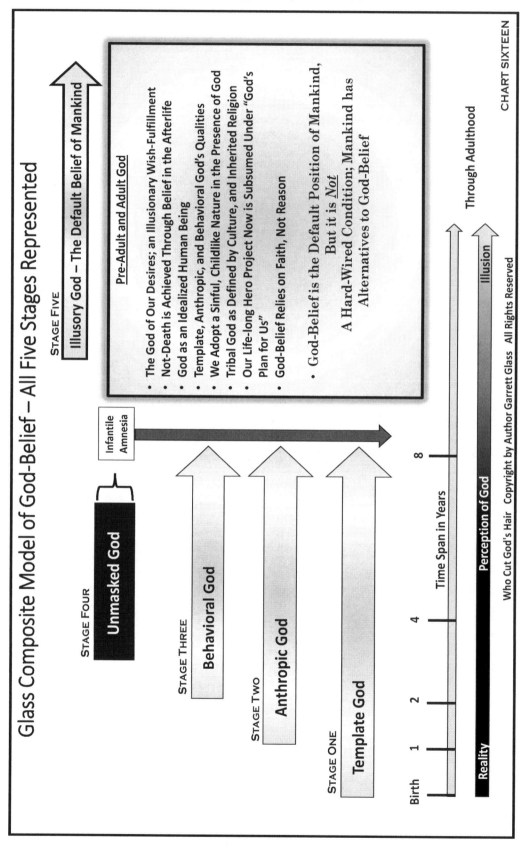

Glass Composite Model of God-Belief – All Five Stages Represented

STAGE FIVE
Illusory God – The Default Belief of Mankind

Through Adulthood

Pre-Adult and Adult God

- The God of Our Desires; an Illusionary Wish-Fulfillment
- Not-Death is Achieved Through Belief in the Afterlife
- God as an Idealized Human Being
- Template, Anthropic, and Behavioral God's Qualities
- We Adopt a Sinful, Childlike Nature in the Presence of God
- Tribal God as Defined by Culture, and Inherited Religion
- Our Life-long Hero Project Now is Subsumed Under "God's Plan for Us"
- God-Belief Relies on Faith, Not Reason

- God-Belief is the Default Position of Mankind, But it is *Not* A Hard-Wired Condition; Mankind has Alternatives to God-Belief

STAGE FOUR
Unmasked God

Infantile Amnesia

STAGE THREE
Behavioral God

STAGE TWO
Anthropic God

STAGE ONE
Template God

Birth   1   2   4   8

Time Span in Years

Reality   **Perception of God**   Illusion

Who Cut God's Hair   Copyright by Author Garrett Glass   All Rights Reserved

CHART SIXTEEN

— 108 —

It is at this point that we come to the end of our model for the development of God-belief in children, because we have reached that moment where belief in God becomes self-perpetuating. The young child is primed to believe in Illusory God, and indeed has the desire to do so and a longing for such a presence in their life now that they have lost all the previous, real Gods as a conscious presence. The parents also have a desire to pass along belief in Illusory God to their children, to provide them with a surrogate parent once they themselves are dead. This process repeats itself generation after generation, reinforced through social and cultural acceptance of God-belief that a young child experiences as something so casually and universally adopted by those all around them that there is never any need to question the material existence of God.

There is no greater social or cultural force at work here than religion, and we are now prepared, with our model of God-belief in hand, to explore religion and the adult world of God-belief in more detail in Part Two of this book. In Part Two, we are going to turn away from science and fact and logic, which have served us well in exploring where God belief arises in the human mind. Instead, we are entering the land of philosophy, of supposition, of subjective rather than objective judgment. In this world of philosophy and its religious counterpart – theology – people are perfectly free to make statements that need not be proved as fact. We are moving away from the realities of the physical world, and entering into the philosophical musings of the metaphysical world – the world beyond the physical.

In this world – especially in the pre-adult phase of our lives – we will meet with yet another manifestation of God – the all-important Tribal God. This is the God which our culture imposes on us, especially as a "gift" to us from our parents, who wish to hand down to us their own religious traditions, and whose motivation is to protect us from inevitable death when they themselves are no longer present in the world. Tribal God is *not* a part of our five-stage model of God-belief, even though our parents or community may introduce us to Tribal God (Allah, Jesus Christ, Krishna or other manifestations of God in Hinduism, and so on) as early as age two when we begin to speak and learn to read. Because there is nothing real about Tribal God, and because Tribal God is illusionary in nature, Tribal God is a component, but an important and often life-long component, of Illusory God.

For the believer in God, our faith in Illusory God and a future existence with him in an afterlife provides us with quite a wonderful world, with numerous compensations, of which the foremost is the promise of not-death. We need not spend too much time exploring these wonders, as there are thousands of religious and spiritual books which do the same. Instead, we will act as impartial observers – as social scientists – probing behind the scenes to understand what motivates adults to continue with their belief in Illusory God throughout their lives, and we will probe the nature of religion as well. Most of the time during this part of the journey, we will use the term God, but occasionally the term Illusory God will be used, to remind us of the true nature of the God that many adults believe and accept in their lives.

First, however, we need to address an important question many people will raise regarding the scientific research underway to discover the origins of God-belief. The question is this: isn't all this research merely revealing how blessed we are that God has given us such a marvelous brain? We investigate this question in the next chapter, titled "Didn't God Give Me My Amygdala?"

PART TWO

THE PHILOSOPHY OF GOD-BELIEF

CHAPTER NINE

Didn't God Give Me My Amygdala?

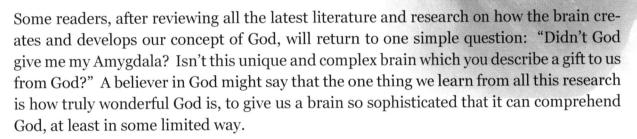

*A casual stroll through the lunatic asylum shows that
faith does not prove anything.*

- Friedrich Nietzsche

Some readers, after reviewing all the latest literature and research on how the brain creates and develops our concept of God, will return to one simple question: "Didn't God give me my Amygdala? Isn't this unique and complex brain which you describe a gift to us from God?" A believer in God might say that the one thing we learn from all this research is how truly wonderful God is, to give us a brain so sophisticated that it can comprehend God, at least in some limited way.

This might seem at first glance to be a chicken vs. egg problem. Which came first: humans who evolved into a creature with such highly-developed cognitive skills that we are able to invent God and give him many human qualities; or God who created humans in order that they may benefit from his love for all mankind? For a person of faith, this is an easy decision: God came first, because God always comes first. For a person of faith, this statement is a testament to their faith, because such a statement requires a precondition – that the person making the statement believes fully in the existence of God in the first place.

Later in this book, we will talk at greater length about faith and its importance to God-belief and religion. For the moment, let us merely state that the question of whether mankind creates God, or whether God creates man, is not as simple as religion and God-belief make it sound. As we saw in Part One, a considerable amount of science now explains how our brain summons God when he is needed. But even more than that, with our knowledge of how the brain works, mankind has taken away from God some of his powers, making him even more remote from the human condition than ever before.

Man's history from primitive times up to the present has been a journey which has forced us to constantly whittle down God, which would be the opposite of God's desire if he made mankind in order to share his universe and his love for us. There are many examples of this, but let us take just two to illustrate the point.

In February of 1504, Christopher Columbus was provisioning his ships in Jamaica, and was dependent on the local Taino Indians for food. The Taino decided to withhold provisions from Columbus, and in thinking how to respond, Columbus turned to an astronomical book he brought with him on the voyage. The *Ephemeris*, written by the astronomer Regiomontanus, charted future solar and lunar eclipses, and Columbus noted that a lunar eclipse was due on the following night. Columbus approached the Taino and declared he would cause the moon to disappear if he did not receive gifts of food. As the eclipse commenced, the Taino were horror-struck, and brought bountiful food to Columbus, begging him to return the moon to the sky. Columbus stayed in his cabin until his charts told him the peak of the eclipse had occurred, at which point he announced that God was pleased with the Taino, and the moon would now return.

From the viewpoint of Columbus' crew, this was a wily use of advanced European knowledge to exploit the ignorance of the Taino Indians. Columbus himself must have believed that this was proof that the Christian God was superior to any pagan god (Columbus was a devout Catholic). To the Taino, however, this was an important milestone in their own history of God- belief. It had now been shown that their god was inferior to the Christian God, and that the Europeans had power to overthrow their gods completely. This was a devastating loss to people who were animists, and who relied on the gods of the forest, the sea, the moon, and other elements of nature to guide their life. You really can't get much closer to a god than to embrace the very tree or animal in which he resides. Columbus' trick made the Taino gods more remote from the Taino people – less personal, and thus less valuable.

Our second example was discussed in Chapter Three, and that is the discovery in the past two decades how to create an intense, spiritual connection between ourselves and the universe, or God if you prefer. These experiences in the past have been reserved only for a select few – for example, someone like Saint Theresa of Avila, a Spanish mystic of the 17th century who claimed to experience visions of Jesus Christ inhabiting her body and allowing her complete union with God. Christianity has always had its mystics and visionaries who have unusual experiences which allow them to join with the "divine." Similarly, Sufism is an Islamic sect that also practices mystic rituals, including reciting repeatedly the 99 names of Allah, thus allowing the supplicant to achieve a trance-like state of Oneness with Allah. Buddhism, Hinduism and other great religions have similar mystic practices which attempt to achieve the "numinous," or an experience of unity with "the Other."

What has been discovered in the past two decades by neuroscientists is that these numinous experiences can be reproduced in laboratory settings by administering psychedelic substances such as psilocybin and LSD. The human brain is seen to be activated in the same manner whether a subject is under the influence of a psychedelic drug, or is engaged in intense meditation, such as a mystic like St. Theresa experienced. Mankind has uncovered the secret of the numinous experience, but we have done more than that. We have *created* the numinous experience, so that under the right circumstances, all can benefit. Mankind has done that which was previously seen as granted only by God, and was accessible only to the few who were worthy of the effort and diligent enough to put in the years of work to achieve Oneness with the universe.

With these discoveries, God has shrunk yet again in his powers. It is true that some of the subjects who have undergone an experience with psilocybin or LSD claim that they have come closer to God, or believe more fully in God's existence. But we are all pre-conditioned from birth to believe in the existence of God, and for everyone who has an induced numinous experience who says they came closer to God, just as many others say they had an intense spiritual experience with no reference to God, or they felt a Oneness with the universe, and not with God in particular. The only consensus everyone can agree upon regarding these induced numinous experiences is that they were created by man, and that the experiences can be verified and reproduced. We know what drugs induce these experiences, what dose works best, and what conditions are most salubrious for a favorable reaction. Mankind has, yet again, taken over one of God's provinces, causing him to shrink in terms of his importance and meaning to us.

It is not that a neuroscientist or medical practitioner is "playing God" when they induce a transcendental experience in a subject. The scientist is using modern technology to create conditions which, in the past, were thought to be gifts given only to certain people by God. Science continues to crack open mysteries that were previously deemed to be expressions of the divine, and which had therefore been placed under the control of religious authorities.

And if, as a person of faith, you still fall back on the belief that all of these discoveries are the result of God's revealing yet another aspect of his miracle of creation to us, and that everything in our brain is the result of his creation, you should then ask why it is that virtually all vertebrates have a basic brain like ours, including an Amygdala, and other elements of the Limbic System which have now been identified as contributing to induced numinous experiences. Why would animals need such capabilities as well? Is God providing mystic visions to lowly lizards?

The simple fact is that we share our Brain Stem and much of our Limbic System with thousands of other species on the planet, and we are therefore nothing more than a biological being like any of these other species. We all have evolved similar primitive brains as a means to survive and reproduce, and reproduction is essential to continue life on this planet.

The real dilemma facing mankind is not whether God created man or man creates God. The real dilemma is, as Ernest Becker explained, the tragedy that we have a mind which can grasp the infinite and comprehend what it must be like to be a god, yet we have a corporeal husk which betrays us minute by minute and drives us ultimately to the grave.

Since it is our mind which can imagine what it must be like to be God, it is our mind which summons up God, and which compels us to worship that which we call God.

There will always be those who insist that no matter how many secrets neuroscience uncovers about the way in which our brain conceives and understands God, God came first, and gave man a brain in order to come to know him better. But there is only one way to reach such a conclusion: *by making a leap of faith.* As there is no practical, scientific, or other real-world evidence that an externalized God exists "out there," the only way for someone to entertain such a belief is through faith, which is an act of erecting mental defenses to maintain God-belief despite the lack of any tangible proof of his existence.

It is also important to state that such a person is not delusional or abnormal in maintaining a belief in God. Belief in God is our default position when it comes to approaching the greatest issues we face in life. We each of us have a neural imprint of a tangible, God-like presence in our lives: Template God. We have an infant experience of heaven, which readily translates into a desire for the afterlife. Mankind has always derived substantial benefit from belief in a deity or deities, not the least of which is the fact that it offers an emotional and intellectual path to not-death.

The purpose of this book is to acknowledge that God-belief is normal and beneficial, though there are costs to God-belief and religion. At the same time, there are other emotional and intellectual paths to not-death that are legitimate and provide similar benefits. In order to understand these alternative paths, we first need to go "behind the curtain" of God-belief, and examine such belief clinically, from a distance, as an outside observer. In this part of the book, we will explore what happens when we summon God. What qualities do we give him? Why is he viewed as the Creator, and if he is not the Creator, how do we explain the existence of the universe? Where does religion come from? Why are there so many different religions, and consequently, so many different versions of God? This is a discussion of importance whether or not you believe in God, because our goal is to explain the journey we are all undertaking. Since most people follow the default path for mankind, which is to believe in God and practice a religion, let us look at these beliefs and practices, to see how they operate, and where they come from.

CHAPTER TEN

Who Is God?

Men always make gods in their own image.
-Xenophanes,   c. 500 B.C.E.

Who Cut God's Hair?

Have you seen God's hair lately? Look again at the cover of this book. The fresco by Michelangelo on the Sistine Chapel ceiling is probably the most famous image of God in Western art. Obviously Michelangelo wanted God to look as ancient as the world God created, but does God really need matted and tangled hair that looks like it was last washed at the creation? Out of decency, we won't mention the mess that is God's navel-length beard.

Someone up in heaven is doing an awfully poor job of cutting God's hair. Of course, this being God, he doesn't need anyone to cut his hair. He could simply blink his eyes and create a head of hair and a beard that would strike awe in the heart of the most accomplished and talented hair stylist. You could also say that this is all Michelangelo's fault. But what is an artist to do? Michelangelo was charged with painting biblical scenes on the Sistine Chapel ceiling and walls, from the creation to the resurrection of Jesus. His source of inspiration was therefore the Bible itself, and when it comes to God, the Bible is very clear in describing what he looks like. He looks exactly like us, because we were made to look exactly like him.

It is admittedly a bit silly talking about God and his haircut, but it gets more serious when we think about why God made us in his image. If he is just like us, he has a brain, heart and lungs, a stomach, a penis, and even a navel. For what purpose does God need these human features? He doesn't eat, he doesn't have sex, and he doesn't need to breathe to survive. A divine and Supreme Being would be a contradiction in terms if he had to put up with the everyday burdens of breathing, eating, and watching his diet so as to avoid heart disease and a premature death.

When it comes to God, there can be no contradiction, since he is perfect. Therefore something is wrong with the story in Genesis. This particular body we have has its uses here on earth. It survives quite comfortably in most earthly habitats, with a little adjustment now and then for clothing and shelter. It thrives off the abundance of food the earth has to offer. We have a wonderful, thinking brain, dexterous fingers with opposable thumbs, and a highly-pleasurable capacity for reproducing our species through sex. Our body seems just right for us here on earth. Why does God need a body just like ours if he lives outside of earth, and anywhere he chooses?

There is no answer to this question if we accept the Bible at face value. If God created us, he would hardly need to encase himself in a human body given all the myriad choices he would have, including the bodies of possible sentient species on other planets. He wouldn't give himself a body with organs and functions that are useless to him. God is perfect, and he would choose instead a form perfect for his omnipotence. There is an answer to this question, however, and an obvious one, if we look at it from another angle. Rather than accept the statement that God made mankind in his own image, what if we assert that **mankind made God in our own image**? Looked at from this point of view, everything makes sense.

If we were going to invent a God, how we would do it? The answer is quite simple. The Template was given to us at birth. *God is our idea of who the perfect human being would be like*. He is all-powerful, all-knowing, always present, all-loving, and our Creator, which are all the qualities we discovered about God in our infancy. A more typical way of saying this would involve this phrase: God is Omnipotent (All-powerful), Omniscient (All-Wise, or All-Knowing), Omnipresent (Always Present), Omnibenevolent (God is Good All the Time), and the Creator of the universe (he who made us).

These are the five universally acknowledged powers or capabilities of God, found to some degree or another across all major religions. They derive from our essential needs as an infant to have an adult figure in our life who is supremely powerful, knows everything, is always available, loves us unconditionally, and brought us into this world. There are more sophisticated qualities of God that we learn about in childhood or even later in adulthood. Some of these are:

God is Heaven

A fundamental premise regarding God is that he is inseparable from heaven. Heaven is his abode, a special part of his creation. Where God travels, heaven travels with him. In fact, God rarely travels, or seemingly has little reason to do so. If he interacts with humans, it is through intermediaries such as angels as messengers, or by implanting dreams or voices in the minds of those with whom he communicates. If humans are allowed to see him, he is normally surrounded by the "Heavenly host" of angels and sainted, deceased humans. It is a peculiarity of heaven that we desire it more than we do God, because heaven is a blissful dwelling that is impersonal, and carries none of God's human traits (such as his tendency toward anger or disciplinary action). Heaven has every physical aspect of our infant experience that we most desire: complete safety; lack of want, pain, and suffering; and the presence of those in our life who love us.

God is Eternal

A hallmark of our experience as infants and children is that we have no concept of the passing of time. We assume, therefore, that our life is never-ending, because we know nothing of death. We therefore associate these qualities with heaven, and with the person of God, in whom we grant immortality. This places God beyond time, and because God is the personification of heaven, this places heaven beyond material space. God and heaven therefore are assumed to exist unfettered in any way by space or time. In modern physics terms, they are not subject to the realities of the space-time continuum, which is to say whether this continuum continues its expansion forever, or as some scientists think, will someday reach a limit and begin a contraction back to a singularity present at the Big Bang, is entirely irrelevant to God's existence. He will carry on eternally, even if ours is just one of millions of universes, because his spatial domain is infinite.

God is External

This is a surprisingly important quality of God that was first explored in depth by the great philosopher Ludwig Feuerbach, in his 1840 classic book *The Essence of Christianity*. To believe in God is to grant him an existence separate from our mind, and outside of our body – he is "out there" somewhere in the cosmos. When we begin as infants to associate externality with our parents or caregivers, largely because our Parietal Lobe has now developed to a degree we can differentiate humans who exist in a separate space, and our Occipital Lobe allows us to visualize God as a human, so too is God given externality. The problem with God's externality is that it is a fiction, but one of the greatest importance. As we will discuss later, God is not literally "out there;" he resides only in our mind. It was Feuerbach's observation that granting God externality, which is a necessary consequence of belief in God, creates a "disunion," a term which implies a psychological rift. We literally carve out a part of our psyche into God's domain, separate from our own personality, which causes us any number of psychological conflicts and crises. This is similar to the process of psychological splitting we discussed in relation to Behavioral God. We will explore the conflicts which arise from the process of disunion, or splitting, in a later chapter.

God is Internal

Within most religions, there is expressed a conviction that God can sometimes be internal by infusing an individual's soul with his wisdom or essence. The soul is an ancient concept, most often described as a spiritual essence which is immortal, and which therefore cannot be destroyed. Plato thought of the soul has having three parts and three locations within the body. The *logos* was the mind, located in the head; the *thymos* was emotion, located in the heart; and *eros* was desire, lust or appetite, located in the stomach. The stoics defined the soul differently:

"You are a little soul carrying about a corpse, as Epictetus used to say."
-Marcus Aurelius

In Stoic philosophy, the soul is the essence of an individual, while the body is nothing but the material shell and unimportant. This view was not accepted by Christianity, because of the promise of the resurrection of the body, but Christians today differ on the nature of the soul, and the extent to which it is equivalent to consciousness. Catholics and many Protestants feel the soul is connected with consciousness, and this consciousness continues to exist after death, when both soul and consciousness meet up in heaven or hell. Seventh-Day Adventists, however, believe the soul is immortal but consciousness converts to unconsciousness and sleep after death. Only at the Last Judgment, in their view, will the sleeper awake and be restored to consciousness. Christian denominations also differ on when, and if, the soul is reunited with the body in the afterlife. They also differ on the meaning and extent of God's interaction with the soul. Most often, Christian dogma argues that a temporary infusion of grace into the soul is a benefit bestowed by God upon believers, through the action of the Holy Spirit.

Atheists have none of these problems and do not worry about these nuances, as they believe there is no soul and no afterlife. That does not mean that atheists can live their lives with disregard to their inner yearnings, especially to the fundamental desire we all have for not-death. God-belief and religion are a means of satisfying the critical psychological and emotional longings we have in our constant struggle for not-death, and religion prefers to lodge these longings in an invisible entity called the soul. Non-believers do not accept the idea of a soul, but it does not exempt them from dealing with these universal longings.

God is Everywhere

God's omnipresence is yet another one of his qualities which causes us a conceptual problem. If God is beyond space and time, and if he is all-powerful, then he has command over space and time. He can be both outside of space, and at the same moment, within our space, our universe. He can also be *of space* – found within the essence of every atom in the universe, which we can describe as a particular philosophy regarding God that is closely associated with Spiritualism and New Age religions. The problem with such concepts is that God as the essence of space is an impersonal being, with whom we have difficulty relating.

Some philosophers have gotten around this difficulty by positing that since God is of space, he must reside within us, within our heart, perhaps, or our soul. But if God is within us, are we not part of God, or are we not Gods ourselves? Does this concept not violate the principle of externality which is one of our requirements for God? Indeed it does, and the way to get around this problem is not to think too heavily about God's omnipresence. Theologians therefore are fond of saying that is God is everywhere, so he is available whenever he is needed, and they leave it at that. In response, having seen the progress that has been made in neuroscience that helps to understand from where God-belief arises, we must simply assert that *of course God is omnipresent; he is a creature of our mind and resident there by definition.* He will always be with us and know our innermost thoughts, because we have dreamt him up. God is omnipresent from our perspective because he is pure thought – our thoughts.

God is Existence Itself

The characteristic of God most beloved by philosophers and intellectuals is God as the "ontological ground of all being." Ontology is the study of existence, so this definition has a self-referencing quality to it, but it means that God exists in his own right, and lends his existence to humans, and to the materiality of the universe, as an act of love. God is therefore the source of existence and of all life. The eminent American theologian David Bentley Hart, in his 2013 book *The Experience of God: Being, Consciousness, and Bliss*, wrote:

> **"He [God] is the source and fullness of all being, the actuality in which all finite things live, move, and have their being, or in which all things hold together; and so he is also the reality that is present in all things as the very act of their existence. God, in short, is not a being but is at once 'beyond being' (in the sense that he transcends the totality of existing things) and also absolute 'Being itself' (in the sense that he is the source and ground of all things)."**
>
> **-David Bentley Hart**

In this construction, God is outside of the universe, and must be so if he is to ground all reality by sharing a portion of his existence with the universe. This definition also implies that God is the Aristotelian "First Mover," that entity which never came into being by itself (God always existed), but which caused the universe to exist (God as Creator of the cosmos). David Bentley Hart has a propensity throughout his book to refer to God as "he" without defining why he does so, but in so doing, he points out the serious weakness in the intellectual definition of God as the ontological ground of all being. This God lacks embodiment; he is not anthropomorphized, and without human characteristics he is remote from our personal concerns. We cannot worship or pray to such a God, which is why the God of the philosophers has always remained an intellectual curiosity, of interest largely to theologians, and not a God one will find of interest to worshipers or even most clergy.

God is All-Holy

This critical characteristic of God derives only partly from our infantile experiences, where we come to know God as all-loving. As is commonly said in some American churches, "God is good all the time; all the time God is good." When we idealize God as the most perfect human imaginable, we take these qualities of perfect love and perpetual goodness to another level altogether, by imparting to God a sacred, or holy nature. An all-holy God, God as the "Holy of Holies" in Biblical terms, exists on a higher plane than we do. As infants, we can tolerate this distance between God and ourselves, because we are born almost entirely dependent on God – a lowly, weak creature indeed. But as an adult, the distance becomes unbearable, because our weakness when compared to an all-holy creation of our own mind, is profound and goes well beyond our physical limitations. Our weakness encompasses moral failures, inadequacies of the spirit, and indulgences of the baser instincts of human nature. We can, if this viewpoint is taken to an extreme (and it often is), view ourselves as vile, loathsome creatures unworthy of God's love. This is where Ludwig Feuerbach's observation about psychological "disunion" comes into play. When we combine God as an all-holy deity, with God as an external entity, we cause a rupture in our psyche. One part of us is purified completely and is externalized to an extent that we do not even view him as part of us; the other part left behind is the dross, the sinful being that must earn God's love. In Christianity, an entire theology of human sacrifice and redemption of all humanity from its inherent sinful nature arose because of this belief that we are inherently unholy to some degree (we must be, because God is all-holy). We will talk about this burden of sin at greater length elsewhere in this book.

God as a Creature of Human Culture – The Tribal God

As infants all of us, and it doesn't matter to whom we were born and where we lived, view God the same way. He is all-powerful, all-knowing, always present, all-loving, and our Creator. Our infantile trust in this God, our dependence on him, and our faith in him, is complete.

We've seen in Part One how Template God evolves into Anthropic and Behavioral God, with good and bad human qualities. Eventually even Behavioral God is overthrown when we face the reality that our parents are flawed individuals with limiting human powers, and they will eventually die and desert us. Worse still, *we* are condemned to die, and the sheer terror of this understanding causes us to repress such thoughts, and bury them deep in our unconscious mind. Now, however, we need God more than ever, and while some children can overcome this urge and grow up not desiring or believing in a God, most seek out Illusory God, who is also the God their parents wish them to acknowledge, honor, and obey. In these middle to late childhood years, in most families the parents require their children to accept the God the parents themselves worship. This is where culture comes into the picture, and where the child, soon to be a young adult, is introduced to his or her Tribal God. It should be iterated that Tribal God is *not* part of the Glass Composite Model of God-Belief, because Tribal God is an aspect, or overlay, of Illusory God. A child may be introduced to their Tribal God as early as age two or three, when language and reading begin to be learned, but there is nothing real about Tribal God. He is the invisible clothing that Illusory God wears.

As an older child, we are primed, and highly susceptible, to being given a new version of God, whose foundational basis is the real God of our infancy and early childhood in all his manifestations, but whose illusionary qualities are about to be interpreted through multiple cultural filters. Our cultural backgrounds are extraordinarily complex, because our culture consists of almost everything that shapes us as humans. Culture is something as basic as our home language, family upbringing, economic and financial status, and educational opportunities. These are the personal aspects of culture, formed in our family setting (which may be an orphanage, or boarding school, or some setting other than our biological family). Culture is also imparted to us through the society in which we live, which dictates relationships among genders, sexual behavioral norms, moral standards, and political and economic structures.

If you are born a male in Pakistan, your family upbringing may dictate that you are raised as a Sunni Muslim, given an education that is focused largely on the Qur'an, and destined to serve ultimately as the patriarch of a family where the young girls are provided very limited educational opportunities, and expected to devote their lives to a husband through childrearing and maintenance of a home.

If you are born a female in The Netherlands, your family may be non-practicing members of what was once the Dutch Reformed Church, but is now subsumed within the Protestant Churches of the Netherlands. As a woman, your educational opportunities will be first-rate and the equal of anything offered your male colleagues. You may choose a career as an oncologist, and decide never to get married or have children. You may be undecided or uninterested in whether God exists, and are almost certainly not interested in anything the Protestant or Catholic Church has to offer.

If you are a female born into a family of Southern Baptists in the U.S., you may decide to send your children to a charter school rather than a public school, because the charter school allows the students to learn about Jesus and to pray during school hours. Your life outside of the family may be time you share with your husband and children in church activities, including Sunday services, Wednesday Bible study, weekend retreats, and summer Bible camp for the children.

All three of these examples would be considered perfectly normal, honorary, and spiritually acceptable within the societies and cultures in which they occur, even though they are completely different approaches to life. In turn, they would be looked at suspiciously by outsiders to the culture, because from one culture to the next, God is so different. Culture is such a dominant force, that God becomes transfigured into an unrecognizably different entity, when looked at by differing cultures.

The transformation becomes so complete, and so compelling within the culture, that it becomes difficult if not impossible to imagine what it must be like to be someone from outside of your own culture. To a Southern Baptist, isn't it odd that a Muslim must pull out a carpet five times a day, face Mecca, and pray to Allah? To the young Sunni Muslim from Pakistan, how disturbing must it seem that a woman can rise to the profession of a doctor, deliberately avoid having a husband or family, and openly proclaim no belief in Allah or any god at all? To the Dutch woman of no particular faith or strong belief in God, isn't it peculiar that a Southern Baptist women would deny herself a meaningful life found in a professional career, and instead spend nearly all her time day after day, organizing her life around a church?

Culture is so all-encompassing, that even God becomes subsumed within it. From one culture to the next, however, there are some commonalities when it comes to defining God.

God as a Birthright

Across cultures, God is viewed as an entitlement, a necessity, or at the very least, something natural to all mankind. This comes for the most part from parents or caregivers, who often feel an obligation to impart belief in God to their children. As a parent, we are acutely aware of our mortality, and ultimate separation from our children. By teaching them about God, we are preparing them for a surrogate parent once we die. Parents who have a religious belief have an added incentive to encourage God-belief in their children, because they can now participate fully within their local church, synagogue or similar religious organizational structure. Even in modern societies where many parents choose not to pass along to their children a religion or a belief in God, it is the unusual or unordinary parents who teach an outright *disbelief* in God. Most parents who are atheists or agnostics prefer to allow their children to decide for themselves when they are adults whether they wish to believe in God or wish to follow a religion. In so doing, even these parents acknowledge that belief in God is some form of a birthright, or at least one path to achieving what we think is our true birthright, which is not-death and the restoration of the heaven we once knew.

Heaven as a Reward

As an adult, for us heaven has ceased to be a surface memory from our infancy. It is more of an impulse or desire that needs to be met. Since it is impossible to fully recreate heaven on earth, our impulse takes two forms: to create as much of heaven on earth as we possibly can, and to obtain a bountiful eternal existence in heaven after our death on earth. Our view of the cycle of life, however, is mistaken. We believe we are born, become educated in morals through religion, try to live our life in a moral way, and if so we are rewarded with eternal presence with God in heaven. This is backwards. The reality of our existence is that we are born immediately into heaven, discover that heaven is an illusion, spend our entire adult life yearning for heaven and God, and then head into oblivion when we die. Only atheists achieve a full understanding and acceptance of the real cycle of life. All others are in thrall to the illusory cycle of life, with its belief in an afterlife and a heaven. Moreover, a person's particular belief in heaven is significantly influenced by their culture. Buddhists believe in multiple heavens as well as places of torment to varying degrees, and the more karma one builds up in life, the higher the level of heaven they will achieve. Karma is something like a currency; it eventually runs out, even if it has gained one entrance into the highest heaven. When it does run out, a Buddhist is returned yet again to an earthly existence, to build up a new karma balance for use in the next afterlife. This is a radically different concept from Christian or Islamic heaven, but what is similar across all these faiths is that heaven is a reward in the afterlife, and it is achieved through a life of good works and devotion to God.

God as a Tribal Totem

In the adult world, when anyone mentions God, it must always be assumed that they are talking about "their God." God is so subsumed within a culture, especially one where religion is important, that God takes on the aspect of a totem, which is a figure in tribal cultures that represents the God or spirit worshiped by that tribe. All of us to some degree or another are a member of one or more tribes, consisting of groups of people with shared bonds or interests. Family constitutes an obvious tribe, but so do our neighbors, our work colleagues, and members of the congregation of the synagogue or church or temple or mosque we may attend. Tribe may also refer to something relatively trivial, such as all those people who support a particular sports team.

Tribes become even more powerful when they overlap. African-Americans in the U.S. have an established Baptist denomination formed many centuries ago that binds those who attend church into a combined racial and religious group, or tribe. Sports teams in the American South are closely connected to local religious denominations (Evangelical, Southern Baptist, Pentecostal, etc.), in that the players pray together, prayers are said publicly before each game, and even baptisms are held on the football or baseball field. In these cases, one's personal identification with God becomes reinforced, because God serves multiple purposes – he provides moral instruction, demands communal worship at church, supports the local sports team, and becomes not just a personal God, but a communal God who is one and the same God, but only to members of the collective tribe.

As we shall see, tribalism, while a tremendous force for binding together a community of believers, is a major impediment to creating sympathy or empathy for those who are not part of the tribe.

The Anthropomorphic, Material God

The last major cultural norm which is common among all tribes is the identification of God as either a material human being, or a being with man-like characteristics. In Genesis, God is assumed to have a corporeal presence, and as such, God is *anthropomorphized*, or turned into a man-like figure. This is not to be confused with the word Anthropic. To be anthropomorphized is to be given human qualities, just as animals in Disney cartoons are capable of talking and having human expressions and emotions. Anthropomorphized God is one of two aspects of Anthropic God; the other is the external existence of God in the outside world.

An essential fact of our human capacity to create God is that he is created in our own image, and he must be, because we cannot imagine any other creature on the earth, now or in the past, which has had our degree of self-awareness and our power to shape our universe. We are, in comparison to all other living forms on earth, truly God-like creatures. Our idealized human being, therefore, as God, must be a physically perfect specimen of ourselves.

God as an idealized human being does not have to be visible to us to the degree that we all know what he looks like. In fact, there are surprisingly few representations of God in Western art, the most famous being that of Michelangelo used as the cover of this book. Most denominations of Islam declare that Allah cannot be represented in his physical form (because he has none), and neither can his Prophet, Muhammad (because this would be a form of idolatry). Allah is nonetheless an idealized human being. In many of the great mosques of the world, the 99 attributes of Allah, as identified in the Qur'an or the Hadith (spiritual writings composed after the formation of the Qur'an), are inscribed in calligraphy on the walls or domes of the mosque. These are human or superhuman qualities: Allah is Exceedingly Beneficent, the All-Holy, the Provider, the Creator, the Almighty, the Giver of Gifts, the All-Merciful, the Loving One, the All-Seeing, the All-Wise, the Resurrector, the Avenger, the Eternal One, and so forth.

The beauty of the anthropomorphic God is that he is a highly flexible creature. He can speak our language fluently, which is essential because he communicates to us through our thoughts all the time. He is entirely identifiable within our culture, and our tribe, because he is a full member of our tribe. We cannot easily imagine him as being a member of any other tribe, and so most often we will describe the God of some other tribe, be it Hindus, Catholics, Wiccans, or even the church down the street, as being a false God.

It is very true that within theological circles, whether it is within the Buddhist, Catholic, Protestant, or many other religions, God is described as a spirit, an invisible, non-corporeal entity who nonetheless remains all-powerful, all-knowing, omnipresent, all-loving, and our Creator. The ordinary worshiper, however, does not identify with this non-corporeal God, which is one of the great contradictions of religion. The ordinary worshiper wants to imagine God with a man-like body, with whom one can have a conversation. This man-like God also takes on the important persona of the Father. God the Father is a figure found to varying degrees in Judaism, Christianity, Paganism, and other religions (though only indirectly in Islam), because he presents such a perfectly natural image for us. He is our surrogate Father, the idealized human parent of our infancy.

Notice that in this discussion of the anthropomorphic God, we use the terms "he" and "him." Modern society is a patriarchal society, with men at the helm of society, be it in the family structure, within business, or in politics. This situation is changing dramatically for the better in many industrialized, developed countries, where women are given equal educational and career opportunities as men, and are now advancing in the political realm as well. We are not at the stage, however, where women are entirely coequal to men, especially in religion. With the exception of Wiccan and some Neo-Pagan or New Age religions, which celebrate the Great Mother Goddess (such as Gaia), modern religion remains entirely patriarchal. God rests on *his* throne; we are all considered children of God the *Father*.

Similarly, there is no question in modern religion, or much of modern society no matter the culture, that God *exists*. God is presumed to be out there, somewhere. He is external to us, but part of our universe, because he speaks to us mentally and takes an active interest in our lives. In Christian terms, he even intervened in human history through the blood sacrifice of his only son, so that we may all obtain eternal happiness with him in heaven. At the same time, God is not *of* this universe, because he created it. He is infinite in dimension, and not subject to time. If this God is too difficult to comprehend, it is because we are said to be lowly creatures in comparison, too feeble in intellect to unlock the mystery that is God.

How true is it that God exists in a real, material sense, or at least in a sense that he can cause our material universe to come into being? How true is it that God exists in a manner external to us, beyond our mind and body? We will explore these important questions next.

CHAPTER ELEVEN

Does God Exist?

I believe in God, only I spell it Nature.

-Frank Lloyd Wright

Breaking Through the Illusion

If God is a creation of our mind, and if he exists only in the context of our thoughts and emotions, he cannot exist outside of our mind, because we are not the creators of the universe. We lack the power to transform our thoughts into a material being greater than all the galaxies which ever existed. If we accept the basic premise that God is a creation of our mind, we must reject the premise that God exists as a material entity, external to our thoughts.

Yet we also said in the previous chapter that materiality is one of God's most fundamental qualities, just as is externality. We invent a God, based on our infant experiences with a real God, and then we impart materiality and externality to this invention, so that we can recreate in our mind, for the rest of our life, the comfort that the God of our infancy still exists – out there, somewhere.

It takes a deliberate, conscious effort to stand back from this process, and comprehend our role in the creation of such an *illusion*. And it is an illusion, in the classic sense of the term, as defined originally by Sigmund Freud. He classified an illusion as a belief inconsistent with reality as it now exists, and which carries within it a wish-fulfillment. Our adult belief in God is premised on a reality which no longer exists, yet carries within it a desire for that reality to be restored. This is our Illusory God, which is in contrast to a *delusion*, which Freud defined as a belief in a reality which is utterly false – which does not now and never could exist in the material world. Belief in God is, in that sense, not a delusion.

Since billions of people entertain the illusion that God exists, what does it take to break through that form of thinking, and to accept that the material, external Illusory God is nothing more than a mirage? What it takes is a conscious use of our powers of reason and logic, so let us put rational thinking to good use, and investigate this question of the existence of God.

The Material and External God

Let's first start with some definitions. The material, external world can be identified as the world around us, outside of our body, and perceived through our mind and our bodily senses of touch, smell, sight, taste, and sound. If you are reading this in print form as a book, or on a computer, what you are holding in your hand exists in the material, external world. You can see it and you can touch it. The chair or sofa or bed you are on is material and external to you. So is the room about you.

What about the bedroom down the hallway? Is that material and external as well? It certainly was the last time you were in it, when your senses could confirm it was material and external. At the moment, though, you cannot see that bedroom. You can only *infer* that it exists and that it is material and external. There is nothing wrong with inferring that something exists – we do it all the time. Sometimes it is quite important for our well-being to be able to infer something exists. If your children are sleeping in that bedroom down the hall, it is vital for your well-being and their safety to infer that the bedroom remains material and external, rather than simply in your imagination.

In most cases, from the moment you sat down to read, nothing has happened in your life to make you assume or believe that your children are not sleeping safely in their bedroom. There have been no loud noises to suggest the bedroom is damaged or does not exist. None of your children has gotten up or even made a noise (at least one that you heard), that suggests they are not sleeping in their bedroom. One or more of them might be awake – that is a circumstance you can understand and accept because it has no doubt happened before. You've walked in on them at night and discovered one or more of them are in bed, awake. You accept that circumstance because they are still safe.

You can continue to sit reading this book because there is no evidence that your children are unsafe or not sleeping comfortably. If there were such evidence, even a minor hint of evidence, you would immediately get up and check on your children. For all of us, our children are about the most important aspect of our life, and their safety ranks far higher than our convenience in reading something.

Notice, however, that your mind is absolutely vital for everything that transpires when doing something as simple as reading a book or sitting comfortably in the knowledge that your children are sleeping safely down the hallway. The sense that you have that you are at this moment sitting in a chair or sofa or on a bed gives you a connection to the material and external world, but it is your brain which is processing the signals brought to it by your senses of touch and sight. Your brain is also vital in making the inference that your children are safe. Your brain takes fact a) nothing has occurred to warn you that the bedroom down the hallway is damaged, with fact b) your children have made no noise that you have heard to suggest they are in any danger, and your brain then makes the inference that your children are sleeping safely in the bedroom down the hall.

You do not know for a literal fact while you are reading your book that your children are sleeping safely. The only way to know this for a literal fact is to move yourself into their bedroom and conduct your reading there. There are occasions when you may want to do just that, when one of your children is sick, for example, and might need immediate care through the night. You want to be by their bedside. It is a comfort to them, and a comfort to you, knowing that you are there in the event of an immediate need. This, one hopes, is an unusual occasion. No parent can spend every waking and sleeping moment of their lives in physical proximity to their children. At some point we all have to accept the fact that we will be physically separate from our children, and our confidence in their safety is based only on inferences we make about where they are and what they are doing in their physical absence from us.

Notice we used the word *confidence* in the last sentence. That word is also found commonly in statistical sciences, when someone might say they speak with 98% confidence that such and such might happen. Used in that sentence, the speaker is conveying the concept that almost all of the time, something might happen. We can also speak with statistical confidence about events not happening. Our lives are filled almost entirely with events that will or will not occur with statistical confidence, we are just not aware of that fact. We can also speak with confidence about the existence or non-existence of things, including God.

We now have identified two frames of reference by which to judge the existence of God in the material and external world. The first is directly through our senses – by touch, by sound or by sight preferably. This is equivalent to being in the bedroom with our children, where we can be assured of their safety and existence immediately through our senses. The second means is by inference, through which we can be assured of God's material existence even though we cannot see him at the moment. We will call that God, Invisible God in our review, but first let us look at whether any direct, physical evidence is available to prove God's existence in the material, external realm.

The Absence of Evidence

In 2003, the United States government predicated the invasion of Iraq on the assumption that Iraq held and might use dangerous weapons of mass destruction. There was no firm proof that Iraq had these weapons, but the U.S. government said that the "absence of evidence does not indicate evidence of absence." This statement sounded pithy and clever, and it's the sort of statement everyone can agree to without thinking carefully about it. Courtroom lawyers like to use statements like this. A prosecuting attorney might challenge an expert witness with the question "You say it is not *probable* that the DNA evidence at the scene of the crime matches with the DNA of the defendant, but it is *possible* that it matches, isn't it?" The witness has to admit it is possible that there is a match – DNA matching techniques are not foolproof – and the prosecuting attorney shuts off discussion and walks away triumphantly expecting the jury to dismiss the witness's testimony altogether. Had the expert witness been allowed to say that the possibility of a DNA match was even smaller than the possibility they would all be struck by lightning in the next minute, the jury would likely pay more attention to the witness and the low probability of a DNA match. It would have been a case where the absence of evidence of a DNA match would be more than enough proof to indicate there likely was no DNA match.

There are many times when the absence of evidence does indeed indicate complete and foolproof evidence of absence. We have no evidence that the sun has ever risen in the west. No such event has ever been recorded in human history, and for good reason. We know why the sun always rises in the east, given the manner in which the earth rotates on its axis and its position to the sun. We can all go to bed tonight with absolute confidence that tomorrow morning the sun will rise in the east.

If someone comes to you and says "just because there is no evidence that the sun has ever risen in the west does not mean it cannot happen," you would think that person was a lunatic. Scientifically, it cannot happen as long as the earth continues to rotate on its axis in its existing orbit around the sun. And the earth will undoubtedly maintain this relationship because the motion of the planets around the sun were set a very long time ago and are not going to reverse themselves.

We do know that billions of years from now the sun will explode and the earth will be incinerated or certainly uninhabitable by our species. It will be a moot point then, from our perspective, whether the sun rises in the east or west because there won't be a sunrise and we won't be around not to witness it. For today, though, we do not spend our time worrying about something that will happen billions of years from now. We take it as a certain fact that *for our purposes* the sun will rise in the east tomorrow and for all succeeding tomorrows. The presence of eternal and perpetual evidence (from the perspective of our species) that the sun rises in the east gives us absolute confidence that tomorrow morning the sun will rise in the east. Similarly, the absence of any evidence whatever that the sun has ever risen in the west is enough evidence for us that the sun tomorrow will not rise in the west.

We can apply such thinking – in fact as creatures of reason whose lives depend on logic and rational thinking – we *must* apply such thinking to the question of the existence of God in the material and external world. There is no evidence that anyone in the existence of mankind has ever met God in a material way, so that his externality to us could be verified. No one has shaken his hand, patted his back and felt flesh and bone, had a person-to-person conversation with him, commented favorably (or even unfavorably) about his haircut, or otherwise interacted with him in a material way. There are no photographs of God, no videos of God and no recordings of his voice. The references in the Bible that God talked directly to Adam and Eve, Moses, or to Jesus have no way of being verified, and must be accepted entirely on faith to be believed (accepted, in other words, through a suspension of rational belief).

If someone says to you that they have had a conversation with God or been to heaven and back, that is not proof of God's existence in a material world. That is merely someone who is telling you what their internal conversations have been like with the God they created in their own mind. Again, that may be a real God to them, but it should not be taken in any way as proof that God exists in the material and external world. If a person persists in the belief that the real God talks to them routinely, or even meets with them, that individual is subject to a professional, psychiatric evaluation, especially if there is evidence the person has other episodes that could be characterized as hallucinations.

If someone challenges you with the statement that while there is no proof that a material, external God exists, you must admit it is certainly *possible* that such a God exists, do not fall for this lawyerly sophistry. The possibility that a material and external God exists is not infinitesimally small; it is zero. There are no provable instances of any human ever meeting the material and external God.

You cannot do an internet search for a photograph of God and find one. In fact, there are only a handful of artistic representations of God, which is surprising considering how important he is in people's lives. This alone should tell us that what is really important in people's lives is the internal, self-created God people imagine exists. People externalize that God and believe soundly he exists in the material and external sense, but the God they pray to and talk with is the God of their imagination.

Are we being unfair to God by holding him to the standard of verifiable evidence of his physicality in the material and external realm? Absolutely not. This is the only standard we can possibly apply, because it is the only way we can be assured of the existence of anything materially. As it is, the Christian God is said to have been physically seen by Adam and Eve in the Garden of Eden, after which he has disappeared forever from human sight. Why is he hiding from us? It is not God who is hiding from us; it is we who have pretended he has banished himself from our sight, and we create this pretense because we have no physical proof of his existence.

All of this reasoning applies as well to a God who is not anthropomorphic, like perhaps a God in the shape of an animal, or a ball of energy. No one has ever witnessed a ball of energy move about with purpose, talk in a human voice, create matter out of nothing, or express love for all of its creation. The same applies to reptilian God, Zeus, Jupiter, or any other representation of God.

We've already discussed how in many parts of your life you are content to operate with an inference of certainty, even on matters of extreme importance, such as the safety of your children. You could not, in fact, live otherwise. Night after night you will go to bed inferring that your children are sleeping safely down the hall. We want to explore next if it is possible to establish a material and external God by inference, through his figurative footprints. We will call him Invisible God – the God you cannot see and who lives down the hallway with the door closed - and we will ask the question whether by examining his "footprints" you can concede that Invisible God has a material and external existence to us.

Invisible God

Many people believe in an externalized God – one who exists in the material and external world – but who is invisible. In many major religions, such as Mormonism, Judaism, and Catholicism, this is indeed the technical theological definition of God: he is a pure Spirit invisible to mankind.

We are used to invisible forces in the world that are profoundly powerful, and which are integral to the functioning of the material world. A good example is gravity, but one can also cite the wind, ocean tides, electromagnetic waves, and so on. The argument is often made that as long as something like gravity exists, which cannot be seen, then God must or at least can exist as well. It is not a requirement that we be able to shake God's hand in order for him to be real, is it?

The answer to this argument, in short, is that – yes – we must be able to shake God's hand in the flesh, or have proof someone else has done so, to agree that he exists in the material and external world. The same rule can be applied to gravity or any other natural force, and all such forces can satisfy the requirement that they are capable of being physically perceived by us in order to establish their real existence.

Here is where our first problem arises in proving that Invisible God is material and external. With our children sleeping down the hallway, we already have decisive physical evidence that they are material and external to us. We see them day after day, and each time we are confirmed in their materiality. Our reliance on inference as to their existence is only temporary, at nighttime, or perhaps when they are outside playing with friends. The ability to confirm materiality and externality is vital to accepting the momentary doubts that may crop up when we are separated physically from our children.

Invisible God cannot meet this standard. We have already established that neither we nor anyone we know of past or present has ever physically met God. This is one reason we call him Invisible God. He has never been physically visible to us, and even though he lives down the hallway we have never been in his room to meet him and we will never do so. We need some other evidence that he is material and external in the room in which he lives. Let us therefore turn to our example of gravity, and see if we can understand better why we accept the existence of gravity despite its invisibility. That may give us an analogy or clue to understanding whether Invisible God has a material and external reality.

Suppose a large crane is standing out in the open, and it is suspending at a height of three stories a grand piano that it can let loose at your command. We explain to you that the crane and the piano are there to demonstrate the physical reality of gravity in the material world. To provide you with absolute, incontrovertible proof that an invisible force such as gravity exists in the material and external world, we ask that you position yourself directly under the grand piano and give the order for its release. Would you do so?

Of course not. Unless you are completely ignorant of the way nature works on planet earth, you know exactly what is going to happen to the grand piano and to yourself as a consequence of its fall. If we asked you to step away from underneath the piano to a place of personal safety, and asked you to order the release of the piano on the understanding that you would have to personally pay for any damage to the instrument, you will still not release the instrument. From that height, the instrument could well be destroyed.

Your knowledge of gravity is the result of your entire human experience dealing with it day by day. Gravity keeps you bound to planet earth. Perhaps a day doesn't go by when something in your hand doesn't fall to the ground, or when you have to pick something up off the ground. It is not just the sum total of these experiences that convinces you of the materiality and externality of gravity – it is the consistency of performance. For items of the same mass, gravity works the same way at all times on the earth's surface. As another example, x-rays have a consistent and deleterious effect on the human body at certain doses, as unfortunately does nuclear radiation. With this knowledge, people think carefully about deliberately exposing themselves to such radiation at high doses, because the results on humans are consistent and with a high enough dose will be fatal.

Can Invisible God provide the same consistent effect on humans as these invisible forces of nature? We assume here that the effect will be *beneficial*, and the emphasis is on the *same, consistent effect*.

Now you might say, wait a minute! Why should Invisible God be subject to the condition that his effects will at all times be beneficial? Isn't the world filled with danger? Cannot catastrophe occur at any time? What about the existence of evil? Hasn't God given man free will, so as to choose between good and evil, thereby determining his eternal destiny?

The question of the existence of misfortune, evil, and good in the world will be dealt with fully in Part Three of this book. For our purposes now, we must point out that these are *religious* questions at stake here, largely because they presuppose the existence of God in the first place, and we are trying to establish this very point – does Invisible God exist in the material and external world? Once we establish the truth or falsity of this statement, we can explore religion and related issues such as the existence of good and evil.

Just as important, we have to be rigorous about our thinking process, and rely on rational and logical analyses and conclusions. We have already rationally concluded that God, if he exists in the material and external world, must be all-powerful, all-wise, and all-loving. In other words, God must be good all the time, and all the time God must be good. Therefore, our condition that Invisible God must provide us with material and external effects that are consistently beneficial to us is a fair and necessary condition by which to judge his existence.

One common example given by believers in God, Invisible or otherwise, is that they know he exists because of the majesty and grandeur of creation, from earth's mountains to the beauty of the galaxies and celestial nebulae. There is absolutely no doubt that certain vistas of nature can be inspiring and lead one to believe that some force beyond the power of humans must have created such beauty.

The problem here is that if that force is God, he does not create such beauty with consistency, and by no means is his creation always beneficial. Nature is a process of constant change, of flux between creation and destruction. A volcanic eruption can be a beautiful thing to behold, but only at a distance. The victims of something as long ago as Vesuvius, to the modern victims of the Mount St. Helens explosion, have a right to question God for his cruel and arbitrary treatment. The same could be said for anyone killed in a hurricane, tsunami, avalanche, sea wreck, cyclone, and so on. Nature can be cruel, and if God is Nature's master, the blame ultimately lies with him.

Suffice to say, if we accept the definition of God as all-powerful, all-wise, and all-loving, he cannot possibly be responsible for Nature at it now is. To put this another way, if Nature is capricious, dangerous, and often deadly to man and all living flora and fauna, then God, invisible or otherwise, cannot exist as its Creator.

It must also be remembered that Nature's laws are sometimes assumed to be evidence of the existence of God. Newton's Laws of Motion and Einstein's Laws of General Relativity are supposedly examples of natural laws created by God and eventually discovered by man. More broadly, most of Nature's workings can be said to be put into being by God and are merely awaiting a time when man can discover and reveal their existence.

The first problem with these beliefs is that Nature did not always exist with such laws. In the cosmic soup that constituted the first hundred millions of years following the Big Bang, there were no stars or planets, and hence no need for something like the laws governing planetary motions and the gravitational interrelationships of the planets. No God created such laws; they evolved as the natural condition of the cosmos evolved. A second problem is that Nature's laws exist side by side and operate in conjunction with Nature's randomness. It is one of Nature's laws that governs earth's rotation around the sun, but since life evolved on earth over one-half billion years ago, however beneficial the constancy of earth's rotation was to such life, there have been five random mass-extinction events during which 50% to 90% of all life on the planet died out. In the most recent event, some 65 million years ago, it appears a comet collided with the earth and made the environment inhospitable for dinosaurs.

By the very nature of these events, they are not consistent. Randomness implies infrequency and inconsistency. These are also the most grievously cruel events imaginable – the entire destruction of whole species. There cannot be an Invisible God directing comets into earth's atmosphere. The act is inherently destructive to life, not beneficent.

It is capricious and not purposeful. We already know that our star is destined to explode and envelop all of earth at some time in the distant future. Now we must believe that at any time, and much sooner, another mass extinction event will occur, taking our species with it in all likelihood. That sort of uncertainty and ultimate destruction is not what a loving God, and a loving Father, would ever have in store for us. Nor would he have perpetrated such wanton death on virtually all creatures multiple times over, long before humans evolved on earth.

For Invisible God to have any meaning, he must perform acts that are consistent and benevolent and that create material and external tracks, or footprints. An example might be the prevention of fatal accidents. If from his closed bedroom down the hallway Invisible God foresaw and stopped all accidents that would otherwise result in a death of a human, that would constitute the evidence of his material existence, because there would be perceived material benefits that could be attributed to him. We would then have confidence in the material and external existence of Invisible God, even though we have never seen him, just as we have confidence in the reality of gravity, because we regularly feel its effects. Invisible God would be a force of Nature that we could absolutely rely on for protection.

Unfortunately we see none of that from Invisible God; we don't even hear him moving about in the bedroom down the hallway. Nature carries on as if he doesn't exist in the real and material realm, and we should carry on in the same fashion.

Whether we look at the direct material and external evidence for God, or the evidence by inference for Invisible God, we cannot deny that the absence of any evidence of a real and material God *is* evidence that a real and material God does not exist. You may go to sleep with absolute certainty tonight that the sun will rise tomorrow morning in the east and it will do so for the rest of your life, and that neither tomorrow nor any other day will a real, flesh and blood God come knocking on your door to have a conversation with you. You may go to sleep with absolute certainty about these things.

Why Does God Seem So Real?

Why is the God created in our mind so real to us? Ultimately our perceptions of the world are lodged in our mind and take on reality merely by summoning our thoughts and memories with our mental capacities. We remember going to the kitchen and getting a drink from the refrigerator because we have done it so many times before. Our brain has memories of the touch of the refrigerator door handle, of touching the container with the drink, of reaching for a glass, of feeling the sensation of cold in our mouth from our first sip, and of tasting the drink by activating certain taste buds.

We don't need to actually hold a drink in our hand or taste it at the moment to know that a drink exists in the material and external world and under the right circumstances can be available to us when we desire it. Similarly, if we are a religious person, we have thought frequently of God, worshipped him in church settings, communicated with him in prayer, discussed him with other people, read about him, sung hymns to him, and knelt to him in adoration. Many of these activities are physical, so we have physical sensations of singing, kneeling, talking, and holding books or papers in our hand which we use to read about him. These are real, material sensations. They create memories for us that suggest in unconscious ways that God is material and external to us.

You might say that the mind is playing a trick on us, by converting the sensations of worshipping God with the idea that he must therefore be real. This is by no means the only trick played on us by our mind. Our whole life is spent perceiving an outer world apart from our brain and body – in other words apart from our senses – that is only a representation of the material world. Perhaps the best example of this is our sense of color. Humans have millions of cone cells in our retinas that are photoreceptors, and which send to our brain messages of the colors they perceive. Human cone cells come in three types that are sensitive to different wavelengths, allowing our brains to recognize three basic colors: red, blue, and green. When we hold a red apple in our hand, our cone cells are stimulated in such a way that the combination of messages sent to our brain transmits to us the color we call red. The fruit itself does not have the color red inherent in its skin; all it has is a molecular structure which absorbs most colors, but reflects back to our eyes in a way that stimulates primarily our red cone cells. As proof that red or "redness" is not inherent in the fruit, other animals perceive an apple as a muddy brown or gray color. Dogs only have two color cone cells in their retinas, so they see the entire world in shades of brown/ black, and blue, and with less sharpness or depth than human vision.

Color and the terms used to describe color are highly dependent on the species perceiving the color, and the structure of their retinas used to organize color, hue and intensity. In the material, external world, the apple only appears to us as red, as an illusion specific to humans. Outside of our mind, red as we know it in an apple does not exist. It is perceptible to us only when certain wavelengths of light excite a specific type of cone cell in our retina. But when that happens, within our mind red appears material and external. It is so real to us that we literally stake our lives on that illusion of reality, by using red in stop lights and stop signs to prevent serious traffic accidents. No person is going to run a red stop light just because they know logically that the light itself does not intrinsically have the quality of red, and that the redness perceived by the driver is just an illusion. Similarly, Illusory God exists only within our mind, but billions of Christians are not willing to bet what they believe to be their immortal soul on the fact that something that seems very real to them is actually an illusion.

When we align this mental illusion with the acts of externalizing and anthropomorphizing God, we have a doubly powerful illusion – a God who exists in human flesh and blood, inhabiting a human body, living somewhere in the world but most likely in the heavens, and so powerful that he is able to communicate with us through thought. Indeed, neuroscientists have established that our Temporal Lobe, which is responsible for hearing, is activated in the same manner whenever a third party is talking with us, *or when we pray to God.* Prayer creates the illusion that a material, external human being is talking with us. With powerful illusions such as these, many people don't need any evidence as to whether someone at some time actually met God. They live their whole lives under the doubly powerful illusion that God exists in the material and external world.

We have shown here that there is no God in a material and external sense. God only exists in our mind, as an illusion, which is why we refer to him in this book as Illusory God. Since this is true, it raises an important question: where did the universe come from? Does not the universe require for its creation an entity exactly like God, all-powerful, all-wise, and existing outside of our mind? Let's explore this question in the next chapter.

CHAPTER TWELVE

Who Created The Universe?

They shall enter the Garden of Eden, where they shall be decked with pearls and bracelets of gold, and arrayed in robes of silk. They shall say: "Praise be to Allah who has taken away all our sorrows from us. Our Lord is forgiving and bountiful in His rewards. Through his grace He has admitted us to the Eternal mansion, where we shall know no toil, no weariness.

– The Qur'an, Sura 35:33

According to Scripture

The most formative event of our childhood is one we cannot recall - that moment or series of moments when we learn the truth about the world: that our parents or caregivers are not the God-like creatures we thought they were, that they will desert us when they die, and that we are doomed to die as well. In this process, we gain devastating knowledge – that there exists life, and there exists death. The first we are granted as if by a miracle, while the last is our ineluctable, dreaded destiny. Is there any particular way to summarize this knowledge – to express it succinctly in a way anyone can understand? The Bible has one of the best descriptions of this knowledge - *it is the knowledge of good and evil.*

When we reach the age of reason, we take a bite figuratively out of the fruit from the Tree of Knowledge of Good and Evil. Recall the conditions that God set down in Genesis regarding the Tree of Knowledge:

> **Now the Lord God had planted a garden in the east, in Eden; and there he put the man he had formed. The Lord God made all kinds of trees grow out of the ground— trees that were pleasing to the eye and good for food. In the middle of the garden were the tree of life and the tree of the knowledge of good and evil...**
>
> **The Lord God took the man and put him in the Garden of Eden to work it and take care of it. And the Lord God commanded the man, "You are free to eat from any tree in the garden; but you must not eat from the tree of the knowledge of good and evil, for when you eat from it you will certainly die.**
>
> **-Genesis 2: 8-17**

The story of Adam and Eve in Genesis is most often interpreted as a creation myth, which it certainly is. But much more than that, it is a coming of age story. The fruit of the tree offers knowledge about the way the world really works. The truth is that the world is not the heavenly abode of our infancy and early childhood. It is a corrupt, material world where death stalks all living things.

We as readers of the story play two roles, simultaneously. We are Adam and Eve, in our infancy and early childhood, brought naked into the world, and planted instantly in the paradise that is Eden, which is our heaven. We are also God, in our adult role as parents, guardians, or caregivers of those innocent youth in our charge who have not yet discovered the ugly truth of the world.

It is often wondered why God places the Tree of Knowledge in the Garden of Eden, knowing full well it is a temptation that Adam and Eve will have to resist for eternity. This seems like a cruel injustice on God's part, but this is a misreading of the story. We, as adults, are God in the Genesis story. We know the truth. We know that knowledge of the truth is inevitable, and that Adam and Eve will not ultimately be able to resist learning the truth. Part of us understands the pain that awaits Adam and Eve, and all children as they approach the age of reason, and if we could, we would withhold that information from them forever. But they cannot stay infants forever, so we wish them to extend their innocence as long as possible, and just as in real life we tell our children to stay away from fire, so in the Adam and Eve story, we, as God, forbid them to taste of the fruit of the Tree of Knowledge.

Sadly, the inevitable happens. The snake entices Eve to take the first bite, and she induces Adam to taste as well. This temptation has for thousands of years been misinterpreted, causing untold misery for women who are wrongly accused of being temptresses because the story is taken literally. There was no Garden of Eden. Adam and Eve are fictional characters who are metaphors for each of us in our youngest years when we truly were in heaven in God's presence. The snake's role in the story is as a representation of the gross materiality of the world we inhabit. The snake slithers along the ground, with no feet; it is the closest animal to the earth – to the reality of the world. The snake is said to deceive Eve, by telling her eating of the fruit will certainly not cause her to die. The snake is not exactly lying. Eve will not die, at least not immediately. The snake knows there is a real world out there waiting inevitably for her. The snake knows the world he inhabits consists of both evil *and good*. The snake also correctly points out that Eve will now know the truth, just as God (we as parents) knows it all too well. Let us see how this plays out:

> **Now the serpent was more crafty than any of the wild animals the Lord God had made. He said to the woman, "Did God really say, 'You must not eat from any tree in the garden'?"**
>
> **The woman said to the serpent, "We may eat fruit from the trees in the garden, but God did say, 'You must not eat fruit from the tree that is in the middle of the garden, and you must not touch it, or you will die.'"**

"You will not certainly die," the serpent said to the woman. "For God knows that when you eat from it your eyes will be opened, and you will be like God, knowing good and evil."

When the woman saw that the fruit of the tree was good for food and pleasing to the eye, and also desirable for gaining wisdom, she took some and ate it. She also gave some to her husband, who was with her, and he ate it. Then the eyes of both of them were opened, and they realized they were naked; so they sewed fig leaves together and made coverings for themselves.

Then the man and his wife heard the sound of the Lord God as he was walking in the garden in the cool of the day, and they hid from the Lord God among the trees of the garden. But the Lord God called to the man, "Where are you?"

He answered, "I heard you in the garden, and I was afraid because I was naked; so I hid."

And he said, "Who told you that you were naked? Have you eaten from the tree that I commanded you not to eat from?"

The man said, "The woman you put here with me—she gave me some fruit from the tree, and I ate it."

Then the Lord God said to the woman, "What is this you have done?"

The woman said, "The serpent deceived me, and I ate."

-Genesis 3: 1-13

Notice how utterly human God is in this part of the story. He walks about the Garden of Eden in "the cool of the day." He is entirely material and external to Adam and Eve. They can see him, converse with him, probably hug him, and certainly hide from him. In fact, God is so human, he is so *us*, that he has no omniscience. He has to call out to Adam – "Where are you?" He does not know that Adam and Eve ate the fruit of the forbidden tree until they admit it to him. Up until this part of the story, we as Adam and Eve are infants, and we as God are parents who are completely human. We are still in heaven.

What happens when Adam and Eve gain the truth? Instantly, they discover their naked-ness. Their childhood is now over. They must clothe themselves to enter the real world. But the story makes clear that the fruit of the tree was "good for food and pleasing to the eye." Eve chose to eat from it because   "it was also desirable for gaining wisdom." The real world has its pleasures, its delights, and its compensations. Still, the truth must now be revealed to Adam and Eve.

"To the woman he [God] said,

'I will make your pains in childbearing very severe; with painful labor you will give birth to children. Your desire will be for your husband, and he will rule over you.'

> **To Adam he said, 'Because you listened to your wife and ate fruit from the tree about which I commanded you, "You must not eat from it," 'Cursed is the ground because of you; through painful toil you will eat food from it all the days of your life. It will produce thorns and thistles for you, and you will eat the plants of the field.**
>
> **By the sweat of your brow you will eat your food until you return to the ground, since from it you were taken; for dust you are and to dust you will return.'"**
>
> **-Genesis 3: 16-19**

Nothing better summarizes the futility, the tragic horror, and the existential cruelty of the human condition, than these poetic and prophetic words from Genesis: "for dust you are and to dust you will return." All the philosophy of anyone who pondered the truth about our fate – from the ancient Greeks, to Nietzsche, Kierkegaard, Sartre, Becker, and so many others – is summarized in these ten words.

It is now that Adam and Eve learn the truth. They are thrust out of Eden – out of heaven – out of their infancy and early childhood – into a world of struggle, strife, pain, and at the end of it all death. Oblivion. Lest there be any doubt about our condemnation, God takes further action:

> **The Lord God made garments of skin for Adam and his wife and clothed them.**
>
> **And the Lord God said, 'The man has now become like one of us, knowing good and evil. He must not be allowed to reach out his hand and take also from the tree of life and eat, and live forever.'**
>
> **So the Lord God banished him from the Garden of Eden to work the ground from which he had been taken.**
>
> **After he drove the man out, he placed on the east side of the Garden of Eden cherubim and a flaming sword flashing back and forth to guard the way to the tree of life.**
>
> **-Genesis 3: 21-24**

There is yet another tree in the Garden of Eden, forbidden to Adam and Eve. It is the Tree of Life, the Tree of Immortality. By denying Adam and Eve access to the Tree of Life, God has confirmed their mortality in the real world that they now will inhabit. The sense of eternal life that they had as infants and young children is now gone forever, never to be recovered. Adam and Eve have lost eternity, and in its place they have become subject to Time, the great destroyer, the ultimate bringer of death.

To emphasize mankind's subjectivity to Time, God summons an angel to guard forever the entrance to Eden, so that Adam and Eve can never return to heaven. The angel carries a flaming sword, which is set "flashing back and forth" for all eternity, like the hands of a metronome – back and forth, forever. God has now set Time in motion, in an inexorable countdown. Time never existed in our infancy and childhood, just as it did not exist in

Eden. But Time exists now, as we become adults. We are constantly aware of its presence, of its demands, and its role in ticking off the seconds of our life. It is at this point that a desire forms in our mind to outrace Time, to cheat it, to obtain not-death. This will be the source of our discussion when we turn to religion and its artifacts of prayer and faith.

The Universality of Creation Myths

For now, it is time for us to say goodbye forever to the God of Genesis. This is the last time God appears in the Bible in fully human form, walking about visible to us, able to communicate with us directly, and operating in our realm on a material level. The God of Genesis is the God of our infancy – our idealized human being, made manifest in the flesh, Template God in particular, but Anthropic God and Behavioral God as well.

Following Genesis, a darkness descends over God, reflecting our own inabilities (and those of the writers of the Bible) to summon up specific memories of God from our infancy. In the very next book of the Bible, Exodus, God appears to Moses first as a voice. He has one of his angels inhabit a burning bush that cannot be extinguished, and speak in God's voice to Moses in that form. To look upon the face of the God of Exodus is to invite death, so awful, or awesome, is God's power and his appearance. Consequently, at one point he instructs Moses to hide in a cleft in a rock; the Lord God will pass by him but Moses is instructed only to look at his back after he has passed by. These are, again, metaphors. The idea that it is death to look upon God's face is a representation of our truth, which is that our heaven died once we left our infancy and early childhood, and the God of our infancy died as well. We cannot visualize him because his countenance is shrouded by our psychological repression and Infantile Amnesia.

When God travels to Mount Sinai to meet with Moses and declare his covenant with the Jewish people, God declares all of Mount Sinai to be sacred ground. Anyone other than Moses, indeed any animal which ventures on to the mountain, shall be struck down dead. God, in other words, is synonymous with heaven. Where he is, so too is heaven. As heaven is a place to which we cannot return in real life, so in Exodus it is death for anyone, other than Moses his servant, to set foot on Mount Sinai.

When God appears before Moses on Mount Sinai, he is in the form of a "dense cloud," a darkness, in other words. This too is a representation of our infant and childhood Gods, lost in our unconscious mind to Infantile Amnesia, and only under unusual and stressful circumstances can any of these Gods be brought forth to our conscious mind. Moses can only hear God's voice; his vision cannot penetrate the darkness, just as we as adults only hear from our childhood Gods through unconscious promptings and instincts. In successive books of the Old Testament, God's interactions with mankind become fewer in number and more remote. He appears to some in their dreams. To the prophet Elijah, God appears in a vision.

When we arrive at the New Testament, the God of the Hebrews is entirely invisible, resident in the Great Temple in Jerusalem, and only the High Priest once a year can dare even to enter God's sanctuary in the Temple – the Holy of Holies. God's appearance at the baptism of Jesus in the river Jordan takes the form of a voice in the sky. At the end of Acts of the Apostles, Jesus ascends to heaven to join his Father. In so doing, Jesus, or the writer of Acts, holds out a critical element of hope to mankind – that we all at the end of our lives shall return to Eden. As God no longer appears to humans in the rest of the New Testament, our own ascent to heaven in the afterlife is our sole possibility of experiencing once again our idealized human being – God. And as seen from the quote at the start of this chapter, Islam acknowledges explicitly that the Garden of Eden is the same as heaven.

The story of Adam and Eve is the second chapter of Genesis. The first chapter is, of course, about God's creation of the Universe. This is God at his most powerful, his most majestic. This is the fundamental creation myth of Christianity, and a story which resonates throughout Western thought. The Genesis story is described as a creation myth. These myths exist in human cultures all around the globe. They are all human-centric myths, even though animals may feature prominently in them.

To cite just one example, the Kintu myth of the Baganda people of East Africa describes the origin of humans. Kintu is the first and sole man living on earth, who was created by God, called Ggulu, who resides in Heaven with his children. These children are allowed to play occasionally on earth with Kintu, and one of them, Nambi, falls in love with Kintu and asks her father if she can marry him. He is very reluctant, but gives his permission only if she will keep the marriage secret from her brother, Walumbe. His name means "one who causes sickness and death." Nambi packs up the things she will need for life on earth with Kintu, including her chicken, but she forgets to include grain to feed the chicken. Against Ggulu's advice, she returns to heaven for the grain and is seen by Walumbe. He follows her to earth to find out what she is doing, and learns of her marriage to Kintu. His presence on earth brings sickness and death to the planet, to Kintu and Nambi (who had been immortal), and to all their descendants.

There are many dozens of similar stories which transcend culture and location. They are similar in that they identify a powerful god or group of gods responsible for creation. They explain how the earth was born, or alternatively how it was created out of chaos. They place a first man or first couple on earth, and challenge them to be obedient to God. Disobedience results in exile from God and the introduction of death and suffering on earth. These myths show a persistent human interest in the origins of the earth and mankind, and an explanation for our difficulties in life, including ultimately our death. They all arise from the same source – the human experience of complete dependence on caregivers during infancy and early childhood, and the shock we all experience when we come to realize the truth of our existence, and the inevitability of suffering, pain, and death. The universality of the human condition in infancy and early childhood explains why there are dozens and dozens of these stories from across the globe, created over thousands of years.

In all of these creation myths, God is a stand-in, a surrogate for ourselves. It is we, as parents, who are telling ourselves these stories, and telling them to our children, preparing them in a gentle way for the inevitable crushing reality of our ephemeral experience on earth. As story tellers, however, we ourselves most often do not recognize the reality of the story we are telling. We honestly believe that the God we are talking about is a material entity, external to ourselves. We do this because a critical part of the story discusses the creation of the universe, which requires a God-like human as creator – in other words, the idealized, all-powerful God of our imagination. Is there something about us that compels us to believe that such an all-powerful being created the universe?

Someone in Our Image as the Only Possible Creator

The most intriguing question mankind ever asks about the real world is: *Who Created the Universe?* The only logical answer to the question is that it is improperly phrased. When you ask about the origins of the universe in this way, you are presupposing that *someone* created the universe, and because we as humans are asking this question, that someone has to be a human being. In fact, that person has to be our idealized human being, or God. The question, asked this way, always leads to the same answer: God, however we define him (because we created him), created the universe.

To understand why the question is inappropriate, and why the answer to the question as it is phrased must always be God (the material and external deity of our imagination), let us first answer the question in a silly way, and see where that gets us.

Let us suppose that a dinosaur created the universe, and specifically, the dinosaur Stegosaurus was and is the creator. Would this hypothesis hold up to rational scrutiny? Stegosaurus is an interesting example because this dinosaur is famous for having a brain the size of a walnut. We suspect that an animal with such limited cranial capacity, limited certainly compared to ours, could not possibly have created the universe, despite the fact that Stegosaurus and most dinosaurs were much larger creatures than we are, and existed on earth for hundreds of millions of years, compared to our four million years (and only about 100,000 years for modern man). Besides, Stegosaurus and all the other dinosaurs have long since died out. How could an animal that no longer exists have created the universe? The Creator of the universe must be many times smarter than us, vastly more powerful, and eternal in the sense that he had to exist before the universe did. These assumptions are forced upon us, especially now that we know the universe is immensely old – about 14 billion of our earthly years old – and infinitely vast. We must therefore conclude that Stegosaurus, an extinct creature, and one with a mental capacity distinctly inferior to our own, could not possibly have created the universe.

In fact, since we are demonstrably the only species on this planet ever to reach our level of intelligence and self-awareness, we are forced to assume that the God who created the universe has to be exactly like us, but infinitely wiser and more powerful than us, and nothing at all like a dinosaur or any other life form that has existed on earth. This assumption exactly fits our definition of God – the idealized human being in every aspect.

This last point is quite important: the Creator of the universe, if there was one, cannot be like any other life form that has existed on earth other than the most perfect human we can imagine, that is - God. This is an assumption we make, and it is so natural to us that we don't even notice it is an assumption. The earliest forms of life on earth are thought to be microbes, invisible to our naked eyes (we weren't here to see them anyway), and like all other forms of life that were or are non-human, we assume they lacked our consciousness and self-awareness. So did plant and insect life, the dinosaurs, birds, mammals, and any other forms of life on earth. We have no evidence that any living form other than human beings has ever had a consciousness and self-awareness *like ours*. Yes, many higher mammals and some birds seem to have a consciousness, self-awareness, and a limited memory. But they do not have these faculties like we do. We have no reason to believe our pet dogs and cats spend their days asking the question "Who made the universe?" If they did, they would presumably also have an ability to talk to us about their observations and conclusions, we could watch them study the cosmos with scientific instruments they invented and built, and we would join them in worship of the Creator of the universe, which in that case could well be a dog or a cat.

In an interesting way, our logic has driven us full circle. We can only imagine that the Creator of the universe is the idealized human of our infancy – all-powerful, all-knowing, always present, all-loving, and our Creator. In other words, the Creator must be God. This is the exact same answer we obtained when we asked the question at the top of this chapter – Who Created the Universe? We concluded that this was a loaded question, forcing us to answer a question about "who" with an answer that can only involve a superior form of human. Whether or not this was a loaded question, we now have two instances where this question of how the universe came to be created leads us to the God of our imagination, Illusory God, who exists only in our mind. Are there any other such examples?

God as the Unmoved Mover

It was Aristotle who first postulated that there must be an "Unmoved Mover," some entity which "moves without being moved." The Unmoved Mover sets the universe in motion, and all things in the universe which move ultimately derive their ability to do so from the Unmoved Mover. The Unmoved Mover, as described in Aristotle's *Metaphysics*, is itself perfect in form, incapable of any change because of its perfection, and able to contemplate, or think, about itself. It exists outside of the heavens, where there are no limitations on space or time, and as such it has no beginning or end.

In his book *De Anima*, Aristotle connected the Unmoved Mover with another of his concepts, that of the Active Intellect. The Active Intellect is the passive human intellect made perfect – consciousness turned into self-awareness, or knowledge put to some creative use. Aristotle was quite vague about these terms, possibly because he himself could not fully work out what he meant by them. Consequently, there have been 2,300 years of debate by philosophers over the nature of the Unmoved Mover and the Active Intellect, plus other related Aristotelian concepts. Christian thinkers in particular found these concepts intriguing, because they pointed to a divine, eternal being as the source of all materiality and motion, indeed all life, in the universe, and that being corresponded very nicely to the monotheistic God of the Bible.

Thomas Aquinas in his *Summa Theologica*, written in the middle of the 13th century, used the Unmoved Mover as one of his five arguments for the existence of God. Aquinas was clear that the motion he was talking about went beyond locomotion – movement from one location to another – and that all natural phenomena involving change (erosion, increases in temperature, etc.), and even life and death, were examples of motion that required some pre-existent, external entity or force to set the process going. He used the term "actuality" to denote such an external entity (Unmoved Mover), acting upon a "potency," or entity in motion, development, thought, birth, and so forth.

Just as important, Aquinas could describe if he wished any number of circumstances whereby a chain of motions is put into place, because the universe functions under cause and effect. What Aquinas could not accept, however, was the possibility that cause and effect could extend for infinity with no beginning or end; that would negate the entire concept of a cause or an effect. There had to be something that set the process in place in the first instance, without which the entire chain of events would collapse. Aquinas was essentially following Aristotle's argument that some unchanging, unmoved, perfect entity had to exist for our world of cause and effect to operate at all. Aquinas called that entity God.

What the Unmoved Mover argument set into motion, at the very least, is what appears to be an infinite number of counterarguments, defenses, objections, debates, and occasional name calling, over whether such a thing as the Unmoved Mover, or God, could or does exist. Much of this discussion has now moved on to the internet, among people with varying degrees of familiarity and passion over the question of God's existence, and the main reason the discussion seems likely to stretch into infinity is because no one can prove through logic or any type of scientific evidence that an Unmoved Mover exists.

Some of the classic counterarguments to the Unmoved Mover hypothesis include the simplistic, but rather important fact that if God is an Unmoved entity, how can he possibility bring about motion in other entities? Further, why must it be the case that cause and effect cannot proceed along infinite lines? It is certainly true that we live in a cause and effect universe, ever since the Big Bang, when a singularity exploded into existence and brought about the space-time continuum. Space proceeded to expand, and time began to exist. From that point on, everything which happened was related to, and at least indirectly the result of, something which occurred earlier. Because we exist in this world, we are *compelled* to think of reality entirely along these lines, which means when we contemplate the singularity, it strikes us as completely illogical – utterly unnatural – that something did not exist before then, to cause the singularity to occur. But is this not a bias on our part? Why could not some other universe exist which does not operate on cause and effect? Why could not the singularity arise out of nothing? "There are more things dreamt of in heaven and earth than in our philosophy," said Hamlet. We cannot say with certainty that an infinite series of causes and effects cannot exist, and therefore the Unmoved Mover hypothesis is not something to be taken as an ironclad rule.

We cannot, for example, say for certain that time operates on a linear scale. Yes, time was set in motion at the Big Bang, and that gives us the impression we are moving forward on a linear scale. But suppose time is actually measured on a circular scale. If we set the Big Bang at midnight, we are proceeding minute by minute, hour by hour, around the hands of a clock. Eventually we will reach midnight, 24 figurative hours later, and the clock will reset with yet another Big Bang. We do not need an Unmoved Mover, a potency, or a God, for a universe of this nature to work.

If we take one step back from the arguments and counterarguments regarding cause and effect, potency and actuality, and so forth, we come to the observation that these phenomena are important only because we say they are. We perceive their importance through our intellect, our capacity to reason, and our self-awareness. We are aware of the passage of time in ways that other creatures on the planet are not. Even creatures of higher intelligence, such as dolphins, the great apes, crows, whales, and elephants, may have a sense of cause and effect, but only as it affects their immediate existence. To repeat the quote we cited earlier from Arnold Schopenhauer: "Animals hear about death for the first time, when they die." Animals cannot contemplate death as we do, or an infinite chain of cause and effect, or ponder the implications of the singularity which resulted in the Big Bang.

We, however, puzzle over these things, and wonder intensely over the existence of a Creator of the universe. Indeed, we seem compelled to think about this question as part of our intensely curious nature, and every time we think about God as the Creator, we bring God into existence. This is so because God is a creation of our minds, born of our infancy and early childhood experiences with God-like creatures who took care of us in an idyllic setting. When we pause for a moment to bring Illusory God forward into our consciousness, he immediately springs into being. It is then, and only then, that God "exists."

For the many millions of years that different forms of life other than human beings existed on earth, not one of them ever asked the question, "Who created the universe?" It was of no interest to them, and it never occurred to them because they did not have the brain, the consciousness, and the self-awareness necessary to ask such a question. Through all that time, and until we came along, God the Creator of the universe did not exist, because there was no life form on the planet to contemplate his existence.

At some point, and we don't know when, our species will head to extinction as will happen to all species on this planet. This is a misfortune, but misfortune is one of the characteristics of the universe. Even planets and stars die. When we disappear from the universe, taking with us our consciousness and self-awareness, God, the presumed Creator of the universe, will disappear too, because there will be no life form in the universe to think about him. At least we presume that is the case – there is always the chance some other sentient forms of life exist in the universe, and all the various sightings of unexplainable aircraft give at least a suggestion, unprovable at this point, that we are not alone in the universe and that we are not even the smartest species in our portion of the galaxy. But if there are other life forms smarter than us, and if they have a concept of God, it will likely be in their shape and capacities, and certainly not ours, because we are demonstrably less intelligent than they are. For them to think that God the Creator exists in human form with a body like ours, including our brain, heart, and lungs, would be absolutely ludicrous, considering how primitive we are in comparison to them. It would be just as ridiculous as our imagining that God has the body and brain of Stegosaurus.

What Science Teaches Us about the Origin of the Universe

We know it is useless for us to ask questions about who created the universe. It is equally ineffectual to ask what created the universe, or why it exists. We do know something about how the universe came about and when that occurred, because some answers to these two questions are currently accepted as part of the Big Bang Theory. That doesn't mean scientists should ignore these questions. There will always be some cosmologists, astrophysicists, and others interested in speculating about the origins of the universe, and whether anything can be identified that brought about the Big Bang. But at this point, there are no definitive, scientific answers as to what happened before the Big Bang. Any theologian or metaphysician who insists they have worked out the answer and comes up with a logical proof as to God's existence as the Creator of the universe, will be offering you nothing more than the type of speculation, interesting though it may be, that came from Thomas Aquinas or Aristotle.

Science, or at least some scientists, can properly address a related matter of equal importance to proving how the universe came to be. They can ask the important question: what does it really matter to us personally, in our limited existence, if we do not have proof as to the origin of the universe, and whether a Creator exists or does not? Very few people bother to look at this question because it is uncomfortable to confront. It is in our nature to wonder about profound questions such as the existence of God and the creation of the universe. There is something about us that demands to know the cause behind something as momentous as the Big Bang.

But if we were able to prove that cause, and even if we could clearly identify the Creator as God, would we not be forced to ask once again, where did God come from? Doesn't our cause and effect universe ultimately mean that we will never be completely satisfied with any answer that suggests an Unmoved Mover, a Prime Cause, or other such supposed final source of all existence?

This possibility was put in similar terms by the great American physicist, Richard P. Feynman. Feynman grew up in a Jewish family in Brooklyn. His father was not a religious person, and he taught him to question everything and seek his own answers to problems rather than rely on other people's solutions. This proved to be excellent advice for a young man interested in science, especially when he found himself one of the youngest physicists involved in the Manhattan Project during World War II, which led to the development of the atomic bomb. Toward the end of his career, in the 1980's, Feynman met with the BBC for a series of television interviews about science and cosmic questions regarding the origin of the universe.

Feynman suggested the universe may be like an onion with an infinite number of layers. Science may uncover the secret to one layer, only to realize that further questions need to be researched and resolved. An infinite series of such mysteries might daunt most of us, but Feynman had an interesting personal perspective on the possibility that there would always be doubt about the origins of the universe, and the existence of God:

> **I can live with doubt and not knowing. I think it is much more interesting to live without knowing, than to have answers which might be wrong...I don't have to know an answer; I don't feel frightened by not knowing things.**
>
> **-Richard P. Feynman**

Feynman is harking back to one of the critical lessons he learned from his father: it is better not to know an answer to a problem, than to entertain in one's mind a false solution. What we know about the universe is that it exists. It is immense, unbounded, and about 14-½ billion years old. We don't know why it came into being and what, if anything, caused it to come into being. It just is what it is. Any attempts we make to identify a Creator of the universe are nothing more than mental games on our part, all of which inevitably lead back to imagining some superhuman as a Creator of the universe. That is all we can imagine, because humanity is how we comprehend the highest form of life on earth, and we must assume a Creator must be a more powerful human being than any we have known. But we cannot as of yet identify scientifically a God as Creator, or any cause behind the existence of the universe. We may never be able to answer these questions, because the universe may indeed be like an onion with infinite layers.

Would our life be any worse if we never received an answer to the question of the origin of the universe? Would our actions on earth be any different? Would we have any less reason to lead a moral life, and to do unto others as we would have them do unto us?

Richard Feynman would certainly say no, our life would be unchanged if the origin of the universe remained a perpetual mystery. We cannot improve on Feynman's answer. Therefore, rather than trouble ourselves incessantly over questions regarding the origin of the universe that we can never answer, we should spend our time ensuring that our ephemeral moment in the universe, both as individuals, and as a species, is as meaningful as possible.

Indeed, Feynman would *encourage us to rejoice in the fact that we do not know and may never know how the universe came into being.* Having such knowledge would close off for us entire fields of speculation, hypothesis, testing, and study. Even more concerning, knowing with absolute certainty that such an entity as God exists would be devastating, and contrary to all that is meaningful in human nature. To understand this better, we turn next to a short tale, called *The Parable of the Middle of Time.*

CHAPTER THIRTEEN

The Parable of the Middle of Time

*I'd like to get away from earth awhile
And then come back to it and begin over.*

-Robert Frost

One day God appeared in the heavens, robed in raiment of glory, and floating atop billowing clouds. His sudden appearance astounded people everywhere, and there was no one who was unaware of God's unannounced arrival, because every television station in the world broadcast his image. "He looks like the Buddha!" said some. "No, he must be Allah," said others. Christians thought he was God the Father, but they would have felt much better if Jesus Christ had been at his right hand. Quietly, and only to themselves, half of the people said "She looks exactly like I thought she would!"

"Where is Jesus?" asked the Christians. "Shouldn't he be here for the End of Time?" "This is not the End of Time," responded God. "We are halfway between when the universe started and when it will end, and since it is the Middle of Time, I decided to appear and see how my creation is doing."

"How do we know you are really God," wondered the people. "I will give you a sign," said God. "I will make the sun disappear from the sky." And so he did. For three days and for three nights, or for what felt like three nights and three nights, the sun was no longer visible. The days began to feel chillier, and the people became frightened.

"Give us back the sun, oh Lord! We believe you truly are God." And so God returned the sun to the sky. The people rejoiced, not merely because the sun had returned to the sky, but because nevermore would they have to listen to debates between atheists and Christians over whether or not God existed.

True to his word, God began to inspect his creation. He would show up walking among the people, suddenly appearing at meals, sitting among legislators in parliaments, or cheering from the stands at football matches. He was very approachable, but he soon discovered crowds were unmanageable. Everyone wanted to talk to him, and most everybody needed something from him – a favor, a boon, a small indulgence of some kind.

"Is there no one here who has everything he wants?" he asked. Some people professed to being completely satisfied with their lives, but God learned even these people always wanted something. They would say to him, "You are God, oh Lord. Cannot you eliminate violence and murder among men on earth?" God thought about this. There was indeed great violence on the earth. Men were vicious to each other when they wanted to be. Of all the changes he could make, God thought this would be one of the most beneficial. So God decided to deactivate the Amygdala in the brains of every person, since the Amygdala was the source of fear and anger in mankind.

Instantly, anger among the people ceased to exist. The people noticed the difference and rejoiced. But then some among them noticed that hunger still plagued mankind, and it seemed to them that a just God would not allow hunger. "You are all-powerful, oh God!" said the people, "But your creation is still flawed. There is hunger upon the face of the earth, and that is because storms and hail and drought often kill our crops. And even when the harvest is bountiful, the work of sowing the seed and harvesting is very tiring, and lasts us through all of our life. A just God, who truly loves his children, would not allow such things as hunger and toil to plague the world."

God thought about this. Indeed it was true that Nature was capricious, and often destroyed the fruits of the harvest. It was also true that the work undertaken by man was grueling and a life-long imposition. Yet these people were beginning to vex him with their constant complaints, and in frustration, God determined to solve mankind's problems with hunger and toil, in hope that his children would finally be satisfied. So God said to the people:

"You are a tiresome people. I will solve the problem of hunger on earth, and of the depredations of Nature, and of the toil you must undergo. Henceforth the earth will produce on its own, and for your benefit, all the food you will ever need. The storms shall be stilled, the ground upon which you walk shall be quieted, and no longer shall you be the victim of capricious Nature." And God bade Nature to cause the earth to yield up all manner of bounty for the benefit of mankind, and to cease any form of natural activity that might harm his people.

The people noticed their new circumstances and rejoiced that they were no longer hungry, they no longer had to work, and that their earthly existence was one of perpetual natural delight. Yet it was still the fate of all mankind that one day, each of them would die. Of all the ills that plagued mankind, death was considered the most fearsome, and the people began to believe that a just God would eliminate this, the ultimate and direst of scourges, from the backs of his beloved children.

So the people said to God, "Oh Lord, most powerful and most beneficent being, we have rejoiced in your loving kindness to us. Yet there is still one thing that gives us great anxiety, and that is the fact that each of us shall inevitably die. You are a great God.

Surely you yourself have conquered death. Why can you not remove this dread fate from the shoulders of your children, whom you dearly love?"

"You know not what you ask," replied God to his people. "You ask to be with me for all eternity. If this is your sincere desire, promise me that this is the last thing you shall ask of me," for God was increasingly distressed that his people had an endless list of wants and desires that only he could satisfy. The people responded, "Truly, God, if you grant us immortality, this is the last thing we shall ask of you."

God thought about this, and in his loving kindness for his people, he decided to grant them this, their final wish. And so God abolished death from the face of the earth, so that all of his children could, like him, live for all eternity. And the people noticed that no one died anymore, and great was their rejoicing indeed, for now all of mankind was immortal, like God.

God said to his people, "I have done all that you have asked of me. I have welcomed you into my eternal presence, so that you may join me in your new home, which I call Eden. I remind you of your promise to me, that you shall never again ask for any favors from me. In return for all I have given you, I ask only one thing: that you cease to be a querulous and irksome people, and instead return to me the love I have bestowed upon you. As proof of your love, I expect each of you to thank me daily for all the blessings I have given mankind."

And the people hearkened unto God, and began praising him daily – nay, hourly for some of them, so delighted were the people to be free from anger, violence, hunger, toil, and most of all - death. In Eden they lived in complete harmony with each other and with Nature. Great were the sounds of rejoicing and love that reached the ears of God, now that his people were present with him in Eden.

After some time, however, the people grumbled that they were tiring of their existence in Eden. Yes, there was no violence, hunger, toil, or death. But Eden was a tedious place, with nothing to do except daily praising God and thanking him. "What sort of existence is this?" they asked among themselves. "In olden days, before God came at the Middle of Time, life was never boring. We always had something to do for ourselves. Now we have nothing to do but praise God, here we are like little infants, helpless and utterly dependent on God, and the worst of it is, our existence in this land of boredom shall never end."

So the people came to God and said, "We have no challenges, God. Is this how you treat your children, whom you claim to love? Why should we spend eternity adoring and praising you when you have condemned us to the most useless of existences? You are a great and powerful God. Return us, oh Lord, to our previous condition, when our lives were filled with anxiety and even terror, yet we were happy because we were truly human, able to make mistakes and yet accomplish great things."

On hearing this, God became excessively wroth with his people, and lost all love for them. Had he not done all that they asked to rid them of their burdens in life, and had they not promised him they would make no more complaints or requests of him? In great anger, God decided there was nothing he could do to satisfy such an ungrateful group of children. So he returned mankind to earth, and then summoned Nature, and bade her to make the rain fall for forty days and forty nights, so that all of mankind might be destroyed. But before the rain began to fall, God's heart softened, and he decided to save just one man and his family, so that mankind could start anew.

God sought out a righteous man who had never complained, and he found such a man in the person of Noah. And God said to Noah, "Oh righteous man, you have never complained like the rest of mankind, and for this I shall spare you and your family from the flood that is to come which will surely destroy all but yourselves. I shall give you the tools necessary to build an ark so that you shall survive the destruction of your fellow men, and you shall bring onto the ark a pair of each animal so that they too can survive the flood."

And so Noah built the ark, and summoned onto the ark a male and female of each animal, and for forty days and forty nights Noah and the ark rode the tempestuous waters and survived the flood. When the rains stopped, Noah came upon the dry earth and offered up a sacrifice to God in thanksgiving for being spared from the flood. And God was greatly pleased with Noah, since it was such a long time since anyone had offered him a sacrifice of any sort.

God said to Noah, "I am greatly pleased with you, my son! I shall restore to you all that was taken from mankind at the request of your foolish neighbors. Henceforth you shall experience anger, and hunger, and toil, and yea, you shall suffer even unto death itself, as befits a being who cannot survive without challenges or the excitements of the mind and the flesh. And I shall take away from your mind any memory of what happened before the flood, so that you shall not know of my visit to earth during the Middle of Time. All you shall know of me is that I Am Who Am. And I shall make your descendants as numberless as the stars in the heavens, and I hereby anoint you with the surname of Nothing, so that all mankind may remember and praise you in the future."

And God did exactly as he promised Noah. Indeed, the people returned to their crude but challenging existence filled with anger, hunger, toil, and the fear of death, but filled as well with all those virtuous qualities of love, compassion, and striving for the good that had once belonged to mankind. All future generations praised Noah Nothing as a righteous man who had survived the flood, and as befits the descendants of Noah Nothing, the people knew nothing of God, other than He Is Who Is, or perhaps He Is Who Is Not. No one was entirely certain any more whether God existed.

And the descendants of Noah Nothing thrived, for indeed, mankind experiences many burdens, and suffers grievous ills, of which death is considered to be the greatest terror of all. Yet there is something more terrible than death, and a burden that, in reality, mankind cannot ultimately survive – and that is incontrovertible evidence that such a being as God truly exists.

CHAPTER FOURTEEN

What is Religion?

My religion is very simple. My religion is kindness.

-Dalai Lama XIV

Regency, Agency, and Community

Religion is created when two or more people come together to discuss and compare their ideas about God, and decide on the proper way to worship him. Any two persons will have the same general understanding of God - he is Illusory God, the combination of the previous manifestations of their infant and early childhood Gods. He is the most perfect human being imaginable. Yet they will have grown up with different cultural filters imposed on their image of God. For example, some will view God as a male, others as a female. Everyone will assume that God spends all his time thinking and speaking in their particular language, since that is the language he uses to communicate to each worshiper. Eventually, two or more people who are of the same culture and wish to worship God in a communal setting, will create their own Tribal God.

At the most basic of differences, Tribal God represents the focus of worship for what we call a religion. A person raised in the Buddhist tradition will have a different Tribal God from another person who was raised as an Anglican Christian. Their cultural filters through which they view God will therefore be substantially different. More typically, the differences a religious person might encounter would be within their own religious community, such as when a High Church Anglican meets with a Low Church Anglican to discuss the esoterica of using vestments and incense during worship services. To the extent a religion wishes to transcend its own culture by evangelizing among other cultures, cultural accommodations must be made. The history of the Catholic Church's expansion in the New World, or across South Pacific cultures, was one of imposing European culture on other cultures, even to the point of demanding that new converts practice sex in a manner different from their own culture. On the other hand, local culture has ways of changing the evangelizing religion and its Tribal God. Religions which expand beyond their own culture are sometimes forced to adjust their worship rituals, or their holy days, to reflect the demands of the local culture. Christianity, for example, adopted the Roman celebration of Saturnalia as the basis for its Christmas holy day.

A lifetime can be spent studying the religions of the world, or even the differences between two seemingly related religious sects within a single denomination. Our purpose here is to focus the discussion on what attracts people to religion, and how this attraction relates to an individual's infancy and early childhood experiences with God in heaven. We will start with some definitions.

Religion: religion is an organization used by a group of people for the purpose of worshiping God. The organization can be massive, global, and highly centralized, such as the Catholic Church, or it can be very fragmented, such as a cult religion with only one leader.

Church: a person's interaction with religion is through a church, synagogue, mosque, temple, shrine, etc. Identification with a church is much more important for a worshiper than identification with a religion.

Functions: religion provides three basic functions. A) Regency is the role played by a church leader (pastor, rabbi, minister, imam) who represents God here on earth, and controls the religious doctrine used by his congregation. B) Agency is the role a church plays in providing social programs for the congregation and local community. C) Community is the powerful sense of belonging to a group that a church provides each of its members. As we discuss religion in this chapter, we will make frequent mention of these functions and describe them in more detail.

Worship: worship is the act of drawing church members closer to the Illusory God of their imagination.

Membership is Not Optional

Worship of God is the most basic of reasons why someone practices a religion. We use the term "practices a religion" rather than "joins a religion" because very few people actually join or choose a religion. In almost all cases, religion is a form of inheritance, passed down from parents to their children, and viewed almost unthinkingly as a generational obligation to fulfill. Religions differ in the intensity with which they concentrate on bringing children into the religion, but the Abrahamic religions (Islam, Christianity, and Judaism) seem particularly aggressive about capturing the attention and devotion of children at the earliest age. The German philosopher Arnold Schopenhauer was the first to comment on this practice, and rather skeptically at that, which was in keeping with his dour and pragmatic philosophical outlook.

"But it is common knowledge that religions don't want conviction, on the basis of reasons, but faith, on the basis of revelation. And the capacity for faith is at its strongest in childhood: which is why religions apply themselves before all else to getting these tender years into their possession. It is in this way, even more than

by threats and stories of miracles, that the doctrines of faith strike roots: for if, in earliest childhood, a man has certain principles and doctrines repeatedly recited to him with abnormal solemnity and with an air of supreme earnestness such as he has never before beheld, and at the same time the possibility of doubt is never so much as touched on, or if it is only in order to describe it as the first step towards eternal perdition, then the impression produced will be so profound that in almost every case the man will be almost incapable of doubting this doctrine as of doubting his own existence, so that hardly one in a thousand will then possess the firmness of mind seriously and honestly to ask himself: is this true?"

-Arnold Schopenhauer

There are a number of important observations in this paragraph: that religion must seek out children because they are vulnerable to the reward and punishment element of religious belief; that religion cannot rely on reason as a tool of persuasion, but must convince believers based on faith; that repetition of the message is critical to completing the indoctrination; and that only 1 in 1,000 children will ever ask the important question: Is this true?

Passing religion on from parent to child certainly accounts for the success of religion over the ages, and it does seem remarkably true that the young adult who breaks away from religion is an exception rather than the norm. But there is more to this than Schopenhauer suggests, starting with the fact that the parents, and the entire family structure, are complicit in bringing children into the "family religion." Children have absolutely no say in the matter, and few children would question the propriety of attending Sunday church, Wednesday Bible study, and summer Bible camp. If they have any objection, it would be that they would prefer to spend their time on something more interesting to them, but children are not allowed to question *why* they must learn about God and religion. Similarly, society accepts absolutely the presumed right of parents or caregivers to raise their children in whatever religious belief they want, which is why highly religious parents are able to pressure school systems and secular society in general if, for example, a teacher appears to contradict the religious beliefs of the parents.

Schopenhauer observes that "the capacity for faith is at its strongest in childhood." Why is this so? What is there about childhood that motivates a child to eagerly accept belief in God and religious worship, beyond the repetition of message that Schopenhauer cites, or the pressure which comes from parents? Note that the age most often selected for religious indoctrination is the age of reason, between years four and eight, and sometimes extending into early teenage years, when Infantile Amnesia has already set in. This is the period in a child's life when they are obligated to begin intensive learning in Torah, or to attend a madrassa to start memorizing the Qur'an, or undergo First Communion, or celebrate their first full-immersion baptism where they dedicate their lives to Jesus as their savior.

Simply put, this is the period when children are exposed to the truth – the dismal knowledge of pain, suffering and death which is the full reality of the human condition. This is the period when all direct memories of this revelation are being suppressed into the unconscious mind, as are any memories of the child's experiences as an infant in heaven with God. The child's emotional and psychological state is such that a substitute God, offering eternal life with him in a substitute heaven, is either eagerly accepted, or at the least unquestioned because it seems a part of the natural order of things. These feelings and impressions are then reinforced over and over with Bible stories or other scriptural revelations which are watered down into easily-digestible fairy tales, and which provide the beginnings of indoctrination into the religious theology the child will need when they reach adulthood.

Throughout this process, an opportunity is never given the child to ask "Why?" or "Is this true?" Not even the adults are allowed to ask such questions. This is not simply a matter of teaching the child or adult the importance of faith; this unquestioning attitude to religious teaching is essential to support the role of the minister, pastor, rabbi, or other leader who is acting as God's Regent on earth. A regent is a substitute for an ailing or incapacitated monarch, and often the regent is the heir apparent to the throne. Religious leaders act as God's Regent on earth, since obviously God is neither a material nor external being, and will never physically show up on earth to lead his people. The Regent must do this for him, and so he assumes both a God-like role, and a God-like authority.

Regents exert the most extraordinary powers over their followers. They take it upon themselves to baptize children into the faith; to force from worshipers their most intimate secrets through confession; to sanctify one of the most important milestones of their life (marriage); to intrude upon their marriage if necessary to ensure that the children of that union are raised in the faith; to threaten them with eternal damnation for disobedience to their dictates (which are disguised as church teachings); to work miracles by conducting faith healings; to demand a tenth of their earnings; and ultimately to be present at their death. To reinforce their image as God's representative on earth, Regents will often resort to saying "God told me," or "God revealed to me," statements which are impossible of contradiction. Whenever a Regent or religious person claims something like "God says," substitute the personal pronoun "I" for "God" to understand what the person really means, as in "I say."

Regents will dress themselves up in splendid robes, much like God would wear, and then sit on an elevated throne, imitative of the throne of God in heaven. In extreme situations, as with cults, Regents can sexually molest children, require that followers forego lifesaving medical treatments, subject their followers to physical beatings, isolate followers from their friends or family who are not members of the cult, "disfellowship" followers from the cult if they object to the demands of the Regent; and even demand that their followers commit suicide if the Regent determines that the End of Time is imminent.

This latter list of atrocities by religious leaders is, most fortunately, a rare exception to the behavior of religious prelates in Western society. It is necessary to point these horrors out because the only protection worshipers have if they find themselves in a dangerous cult is intervention by secular society, and often this intervention comes too late because all religions, at least in the U.S., are given great leeway to practice their worship in secret and with no secular oversight. The other important point to make is that even the milder forms of Regent behavior, such as hearing confessions or threatening someone with eternal damnation or excommunication, all derive from the extraordinary claim that the Regent speaks for God and acts entirely on his behalf. This is a claim secular society is not permitted to contradict, because religious belief is considered personal and intimate and not subject to government control. Except in cults or congregations dominated by a single preacher, regents do not act based on their personal and intimate beliefs. They act on behalf of an institution or organization, and in many churches that are not hierarchical and not subject to some higher ecclesiastical authority, the Regent's word cannot be appealed or overturned. In this sense, Regents hold one of the most powerful positions in society, with no secular oversight whatever.

It is the Regent who enforces doctrinal purity, by maintaining strict control over the religious message. It would be an extraordinary act of insolence, for example, for a member of a congregation to stand up during a sermon and challenge the words of the Regent. Bible study groups, to give another example, allow for discussion of Bible passages and their pertinence to modern society, but in the end, the group is not allowed to stray from an orthodox interpretation of the scripture as the Regent sees it. Depending on the religious denomination, Regents may themselves be constricted in interpreting scripture. Many large, traditional denominations within Christianity have a hierarchy of bishops or elders who enforce church doctrine down to ministers within churches. The Catholic Church is the most extreme example of this, with enforcement of the Magisterium globally resting ultimately in the hands of a single individual, the Pope. At the other end of this extreme are the loosely-organized churches operating within the American Evangelical, Fundamentalist, Pentecostal, and related traditions. In some individual churches, the religious credo is simply stated as "Jesus Saves! Religion Kills!" meaning on the surface that salvation is between the believer and Jesus directly, but of course meaning in practice that the pastor as Regent ultimately determines how this relationship is to be conducted.

Two processes are therefore at work which bind a member of a church to that religion. One, the member almost certainly has no choice in their beliefs in God and a particular religion; these were given to them as a child. Second, within the church structure, the Regent acts as enforcer of doctrine and belief. Most often this enforcement is done with a gentle, if not loving hand; the overwhelming number of pastors and ministers are sincere about their beliefs and respectful of the powers that are inherent in their role as Regent. It would be an unordinary situation where a member of the church is losing their faith, or is vocally opposed to the doctrines of their church, that would force the Regent to quietly invite them to leave and seek out a different church or denomination. But if these are unusual situations, something else must be binding members to their church and religion. What is this additional force?

The Psychological Imperatives of Belief

Belief in God entails the psychological process of "disunion," whereby the believer carves out figuratively a portion of their personality that is all-holy, and invests that persona in an illusionary God of perfection. This leaves the believer with an intact portion that is perpetually less than holy. This phenomenon is present in the major Asian religions, such as Hinduism and Buddhism, where the believer is obligated to work in this life to store up favor (karma) so as to allow one's imperfect self to achieve a higher plane of existence in the next life. The Abrahamic religions put a special emphasis on the weak nature of the believer, calling it sinful, and enumerating the many sins an individual can perpetrate as a consequence of their inferior moral inner being.

Sin is a near-obsession in the Abrahamic religions, and is emphasized from the very earliest scriptures in the Bible. Most Christians do not realize that the Ten Commandments given by God to Moses were a very small part of a much larger list of prohibitions on behavior, or instructions on proper behavior. The penalties were extremely severe, often involving death for offenses today that we would consider quite minor, such as a youth talking back to their parents. Why the ancient Hebrews should be so thoroughly concerned about sin has been a matter for scholarly debate, but the consequence for all the Abrahamic religions has been to place sin as a central focus of religious behavior.

The nature of a sin, and the severity of that sin in comparison to others, defines the degree of separation from God that is the consequence of the offense. Clearly within Christian thinking the Ten Commandments rank among the most serious of orders handed down by God, and it is difficult to argue that murder, lying, and thievery are grave offenses against the human victim, not just God. But it is also clear, simply by the order in which the Ten Commandments were delineated, that offenses against man (such as murder, lying and thievery) rank somewhat less in severity than offenses against God.

We learn in the first five Commandments that God is jealous of other gods, and will tolerate no Hebrew man or woman who gives worship to any God but Yahweh. He demands animal sacrifices on a certain day of the week, and he expects his Sabbath, or day of rest, to be honored. Most of all, he despises blasphemy, which can consist of mocking God, declaring that the there are other gods beside Yahweh, or denying that Yahweh exists or is a God.

And why shouldn't God be concerned about blasphemy? In modern terms, and in a very general way, atheism can be considered a form of blasphemy, as it results in a denial of God's very existence. Once someone denies God's existence, they are free of the psychological disunion which is the consequence of belief in God. When that disunion disappears – when someone no longer is plagued by the burden of irremediable, inborn sin – the sense of perpetual guilt which results from a sinful nature disappears. The non-believer is liberated from religious guilt.

Whoever wrote the Ten Commandments, with their special emphasis on believing in and obeying God, had a brilliant insight into the human condition. They understood that belief in God, and its necessary burden of sin placed upon the believer, can be imposed on someone as soon as they leave childhood and come to terms with death and the discoveries encompassed within the truth of their new existence. The first discovery is that having left infancy and early childhood, believers know now that they are separated from God. Second, they realize they have been forced out of heaven forever. Third, they discover they are weak and undeserving creatures, compared with their earlier condition in heaven, when they existed in a state of dependence on God, but without sin. Fourth, the world they now inhabit fills them with fear of evil, and the ultimate reality of their death.

From all this, they feel the separation from God and from heaven. They *feel* it, rather than think about it purposely, because it lays as an emotional memory deep in their unconscious, motivating a perpetual longing for what was lost from their earliest years. This longing is precisely what the Ten Commandments exploits. In the Bible, God is positioned as a material and external entity, who wishes to be reunited with his children, but cannot do so because of their lowly, depraved, and profoundly sinful nature. They must overcome this nature in their daily life, and stay within the rules he has laid out for them in the Ten Commandments, and which will be conveniently interpreted for them by his Regents (who in the Old Testament were the priests selected only from the tribe of Levi). Only then will the believer be reunited with God, in the afterlife.

The believer is now put in the position of accepting intuitively and without question that they are morally weak. Their only hope of obtaining their desired reunion with God is by obeying the Ten Commandments, the first half of which have to do with obedience to God, and the last five with common sense standards (thou shalt not kill) that were already enshrined in written laws such as the Code of Hammurabi long before the Bible was written. These are standards that are essential for the proper functioning of human society, and while they have a moral component, it does not take a religious upbringing or belief to come to the conclusion that if murder, for example, were not forbidden in civil society by law, barbarism would result.

The Ten Commandments were completely superfluous to the functioning of Hebrew civil society 2,500 years ago. They were essential only to the extent the priestly caste assumed the role of civic leaders, and of course to the extent these priests needed to enforce religious dogma. In the same manner, the Ten Commandments are unnecessary for the functioning of modern civil society. What they have accomplished, unfortunately, is to impose a form of moral atrophy on Christian believers, who are taught to assume that all moral guidance has come down from heaven in the form of the Ten Commandments. Moreover, only God's Regents, speaking from the pulpit, may interpret morality or ethics. This is why Christian believers are so shocked when they meet an atheist. How can these atheists prevent themselves from murdering someone when they don't believe in the Ten Commandments? This is a classic case of psychological projection. The believer is projecting on to the atheist their *own* deepest anxieties about what would happen to *them* if the Ten Commandments no longer existed. Their cocoon of moral certitude would collapse about them, along with the whole edifice of sin, guilt, repentance, and atonement. How could they possibly survive, when all moral guidance has come not from their own thinking about morals and ethics, but from the dictates of others who claim to be moral experts?

Perhaps the reason Christians think that atheists live a life of sin since they have no moral standards to guide them, is because a Christian isn't dealing just with the guilt of ordinary sin, or even of a grave sin as found in the Ten Commandments. Christians bear the burden of something far weightier, and that is the guilt derived from *Original Sin.*

The concept of Original Sin is one of the most formidable guilt doctrines ever constructed. Original Sin refers to the disobedience of Adam and Eve when they went against God's commandment not to eat of the fruit of the Tree of Knowledge of Good and Evil. We ourselves know now that this was a just a story, a coming of age myth. But the early Christians, St. Paul in particular, took the story as literal, historical fact, with the gravest of consequences for women in particular. St. Paul set the foundation for the doctrine of Original Sin, which was then expanded upon by men such as St. Irenaeus and St. Augustine. Under this doctrine, the singular consequence for all of us, men and women, and even children, is that Original Sin, over which we have no possible control, becomes our responsibility, our fault, from the moment of our birth. Original Sin is like the guilt inherent in the Ten Commandments, but on hyper drive. It condemns all men to separation from God and his love during their time on earth.

Original Sin is also the greatest of all psychological entrapments. The doctrine asserts to the believer that it was wrong for them to learn the truth about death and the fate that is in store for them. The suggestion behind the doctrine is that they would have been better off staying as young children, never eating of the fruit of the Tree of Knowledge of Good and Evil. They should have never grown up. Unfortunately, they did grow up and learn the truth, and therefore they must be punished for growing up. Their desire, therefore, is to say throughout their life, to their Regents, and to the God for whom he speaks, "I'm sorry for growing up. I didn't mean to. I will be a good little boy/girl from now on!" This inner desire not to grow up, and not to face the reality of their existence causes Christians to adopt the stance of a child as part of their religious identity. Their relationship to their God and to their Regent is one of a child to the father; indeed within Christianity Regents are often call Father, or Padre, or Pére. This is yet another reason why Christians act with profound obedience in the presence of their Regent, why sexual abuse by a Regent of a congregant is such a horrifying crime, and why Christians when face-to-face with an atheist, are genuinely shocked that someone can live without the rules that they, as children, must obey. In those moments, they are displaying the childlike aspect of the psychological burden they carry because of Original Sin.

The psychological burden becomes almost overwhelming. As believers we are only human, and none of us can possibly live a life of such perfection that we do not lie, do not get angry, do not get jealous, and do not succumb to lust. This is what it means to be human, but in Christian terms, this is what it means to be a sinner, and Christianity requires that our sins need to be admitted, we need to achieve honest repentance and regret for these sins, and we then need to seek God's forgiveness, which will *always* be forthcoming if we sincerely follow the three steps just outlined, because God in his perfection is all-merciful.

But try as we might, we sin again, and the process must be repeated, incessantly, for the entirety of our lives, not because we in truth are sinners, but because we are humans who have seen it necessary to seek comfort in a belief in God. This endless cycle of sin, repentance, forgiveness, further sin, and so forth, is both tiring and trying, leaving us in a state of perpetual doubt as to our self-worth.

Why do we put up with this psychological burden? The answer is quite evident – the promise of the afterlife! It is the shiny bauble held out to us as our reward at the end of the journey of life. If you have at all been involved in religion, and most of us have, you understand the power of this promise. A promise of eternity with God in heaven is the one thing that can assuage the pain that we inherited when we learned the truth of our existence.

The truth is repressed to our unconscious level, so that we do not think about it directly unless actually confronted with our imminent death or that of someone we love. The promise of the afterlife, on the other hand, is brought to our conscious self through religion, and is therefore frequently on the surface of our emotions. You might say to yourself, how seriously traumatic can this pain be which is talked about so frequently in this book? Most of us are never aware of this pain caused by our knowledge of the truth. But if the existential anguish wasn't always there below the surface of our consciousness, motivating us in thought and in deed in unseen ways on a regular basis, then the promise of an afterlife would be just words. But it is not just words! It is accepted fervently, without question, by billions of people following an enormous number of religions, and it has been a support for humans for thousands of years. As we shall see, it is not just the religious who are immediately susceptible to the pull of this promise. The secular who have no belief in God and no desire to follow a religion seek the equivalent of the afterlife in the form of immortality through the possibilities of selfless service to others. We are all seeking not-death, in some form or another.

What religion offers is not-death for all who believe, if they lead as blameless a life as possible. Moreover, even in Christianity it is possible to build up a form of karma on earth, which makes the afterlife easier to achieve. It is this propensity to do good that is a hallmark of all the great religions.

Do Unto Others – The Importance of Altruism

A century before Jesus Christ began his ministry, a rethinking of Jewish scripture was underway in Jerusalem. The Ten Commandments, and indeed the entirety of what became known as Mosaic Law, or the laws passed down to Moses by God, were "modernized" in the sense that Jewish teachers tried to winnow down what was truly important from many dozens of regulations and commandments. The greatest of these teachers, Hillel the Elder, was asked around the year 70 BCE to summarize all of the Torah, the first five books of the Old Testament, while standing on one foot. He met this flippant question by answering, "That which is detestable to you, do not do to others. That is the whole of the Torah." Perhaps Hillel was familiar with a saying of Confucius, written four hundred years earlier: "Do not impose on others what you do not wish for yourself." There was some communication and trade between the Roman Republic of the time and the developing Chinese state, so it is not impossible that Hillel the Elder had heard of Chinese philosophy. It is also possible that these two worlds were moving on a similar path of development.

One hundred years later, Jesus Christ was asked which was the greatest commandment. "Do to others what you would have them do to you," was his response. (Matthew 7:12) Jesus was very well versed in Jewish scripture, and no doubt knew of Hillel's answer to that question, which today we refer to as the Golden Rule. Hillel's advice, as with that of Confucius, was passive – *do not do* unto others. Jesus' advice was active – *do to* others. There is some degree of difference in these responses, and Christianity has been influenced by that difference, in that the practice of charity is a direct response to Jesus' teaching. But charitable giving was also an important element of Judaism in Jesus' time, as it very much is today. One of the five basic pillars of Islamic teaching is the obligation to give alms to the poor.

Altruism, which is the practice of making sacrifices for others with no benefit for oneself, or which may harm oneself, is a principal feature of all great religions. The focus can differ quite significantly, however.

Jainism practices the most extensive form of altruism: all living things are to be respected, as all sentient life is sacred. As was said in the *Yogashastra*, around 500 BCE:

> **This is the quintessence of wisdom; not to kill anything. All breathing, existing, living sentient creatures should not be slain, nor treated with violence, nor abused, nor tormented, nor driven away. This is the pure unchangeable Law.**
>
> **-Yogashastra**

Buddhism practices the most intimate form of altruism. Said Gautama Buddha,

> **We are shaped by our thoughts; we become what we think. When the mind is pure, joy follows like a shadow that never leaves.**
>
> **-Gautama Buddha**

Christianity preaches the most extreme form of altruism: the sacrifice of one's life for others. Said Jesus to his disciples:

Greater love has no one than this: to lay down one's life for one's friends.
-John 15:13

Altruism is an instinct and an inborn characteristic shared by all humans, as a result of millions of years of development. It significantly predates religion and would remain a human impulse if religion completely disappeared from modern society. Why, then, is it such an important part of religious doctrine and practice?

The answer to this question has to do with the role religion plays as God's Agent on earth. If a minister, imam, monk, or nun can act as God's Regent, a community of worshipers can act as God's Agent, performing deeds on his behalf here in the earthly realm. Agency is an essential element of religion because it reaffirms the existence of both God and heaven. By engaging in charitable works, religion recreates as best as possible the world of heavenly perfection we enjoyed in our infancy, in the presence of God. Religion appeals to our unconscious understanding that such a heavenly existence *was real*, that it *can be real again* to some degree here on earth, and that it *shall be fully restored* in the afterlife. By making such a powerful appeal to our inner understanding and belief in a God-like figure and a heaven, religion validates its own purpose and confirms the existence of the very God it worships.

Religious charity extends well beyond merely giving money to the poor. Religious charity often reaches into the foundations of society, acting as a critical support for social well-being. This tradition can best be seen in the development of Christianity, which as a religious discipline in its early years of existence offered its believers something society at large could not. The Romans, in conquering the entirety of the lands surrounding the Mediterranean Sea, were excellent at providing social infrastructure (roads, aqueducts, law courts, theaters, etc.), but they did very little in providing social services. One of the benefits of becoming a Christian, even as early as the 1st century, was that believers had access to communal help in the form of food, shelter, and clothing in times of emergency. More important, on an ongoing basis the community provided care for the young, sick, orphans and the elderly. While such care was by no means guaranteed and depended upon the community's resources and stability, the explosion of Christianity across the empire in the 1st century was the result, in no small way, of believers receiving critical social help that was unavailable otherwise.

By the collapse of the Roman Empire in the early 5th century, Christianity was able to replace the political governance structure with its own ecclesiastical institutions, while at the same time expand basic social services, especially at times of extreme famine, plague or war. Medieval Europe represented the pinnacle of Christian temporal and spiritual power: the church was involved in farming; charitable food distribution; education;

medical care through monks, hospitals and hospice care; homes for orphans and the elderly; emergency relief in time of famine; and even entertainment through village fairs and church festivals. The Catholic Church truly was God's Agent on earth.

This model broke down with the rise of a merchant class and the development of the nation state, which slowly began to take over social services from a church that had been torn apart by schism (Catholicism vs. Protestantism). The role of religion as God's Agent began to decline from this point, and the decline accelerated steeply in the 19th and early 20th centuries as Western governments began to provide or subsidize public education, state hospitals, nursing homes, retirement benefits, and orphanages (through foster care programs). The decline of religious practice and belief in God corresponded roughly to the expansion of cradle-to-crave welfare programs by government. The countries in Europe, for example, with the most extensive welfare systems today, tend to have the lowest church attendance and the greatest number of non-believers as a percentage of the population. This should be no surprise, since religion in these countries, through the loss of the Agency function, no longer is seen as important when it comes to care for the sick or elderly, emergency assistance, and most especially, education of the young. Without access to youth for purposes of education in religious belief (indoctrination, in other words), religion slowly atrophies.

It is no surprise that in the United States, religion is extremely suspicious of any expansion of government welfare programs, and often finds itself in alliance with conservative political forces which are also opposed to "big government." Most religious authorities intuitively understand that the Agency function of religion is vital to their survival, because any religion is ultimately judged as creditable and worthy based on its ability to teach its followers empathy, ideally empathy for all living things on this planet. Remove the ability of religion to provide societal programs with a charitable and altruistic purpose, and it becomes more difficult for religion, through its institutions, to argue that empathy (do unto others, etc.) is a prime motivating element of the religion.

In these situations, religion turns inward, and begins to focus on itself – its function as a Community. In providing a community, religion responds to yet another basic human impulse – our desire to belong to something bigger than ourselves, be it a cause or a group with similar values and activities. Yet community can be exclusionary, and this is where religion, if not careful, can hasten its decline in society.

Community vs. Tribalism

For nearly thirty years, starting from 1970, Northern Ireland was wracked by political and religious strife, known now as the Troubles. On one side of this warfare, were the political unionists, who believed Northern Ireland should remain as part of the United Kingdom. The political unionists were Protestants who aligned themselves with the Church of England (i.e., the Anglican Church). On the other side were Irish nationalists, who believed Northern Ireland should be independent of the U.K. They aligned with the Catholic Church, and with Irish sensibilities in general. The Troubles brought about the deaths of over 3,000 people, by murders, assassinations, and bombings, which were perpetrated in London and elsewhere outside of Northern Ireland, in an attempt to force the U.K. government to abandon its political claims on Northern Ireland. In the end, the Troubles came to a halt with a ceasefire agreement that included political reforms which granted Northern Ireland considerable powers of self-government, independent from the U.K. Parliament in London.

The Troubles in Northern Ireland were a recent example of religious tribalism resulting in violence through sectarian (i.e. religious) warfare. The Troubles were by no means a religious war, but religion was a significant aggravating force in setting one group of people against another. A more recent example of an outright religious war is the ongoing violence that plagues so many countries in the Middle East. This is violence which stems ultimately from a struggle for political and religious dominance between Sunni Muslims and Shi'ite Muslims. The center of Sunni identity in Islam is Saudi Arabia, as the guardian of Islam's most sacred shrines in Mecca and Medina. The center of Shi'ite identity in Islam is Iran. The religious struggle between the two principal Muslim sects – Sunni and Shi'ite Islam - automatically takes on a geopolitical nature.

What is it about religion that causes it to be the focal point, or the aggravating element, of civil and international wars? There are two causes. The first is that religion automatically defines God through various cultural filters, which then shape not only the definition of God into an idealized human understandable through that culture, but which also shape the worship practices of that religion, so that they mesh well with the local culture of the believers. Unfortunately, local cultures can clash, with the disruption sometimes turning violent, and when that happens, religion is dragged into the conflict almost immediately, because religion has the ability, since it speaks for God, to add legitimacy to each side of the conflict. It was Abraham Lincoln's observation during the American Civil War that North and South "both read the same Bible and pray to the same God, and each invokes His aid against the other." God could not be on the side of both, and either one religion was wrong, or they both were.

Second, religion by its very purpose creates a sense of community, because worship is more appealing when it is a communal experience. The Community function of religion is a very powerful attraction for believers; we humans have a natural affinity for people like us, starting with something as fundamental as those who speak our language, live in our neighborhoods, and – as much as we don't wish to admit it – are of the same race. Religion adds an important additional dimension of shared identity – a religious community believes in the same God and the same heaven, and if we are a member of this community, they will be the sort of people we shall expect to meet in heaven.

The separating line between a community and a tribe is very difficult to describe, but we can say that a community has no difficulty engaging other communities, while a tribe tends to emphasize the uniqueness and superiority of itself in comparison to others. Religious sects are, for most part, created by and defined by their tribal characteristics, notably on religious doctrinal matters. One religious sect may make the centerpiece of its worship the handling of venomous snakes, because the Gospel of Mark specifically mentions that those with faith shall be able to overcome the bite of a viper. Another religious sect, in the same community, may emphasize the healing of sickness through the hands of the minister, working miracles on behalf of Jesus.

Religious dogma is just one significant way worship communities distinguish themselves from other communities, which may be competitors for congregants. Culture is just as significant, if not more so, because there are many different ways culture can be delineated. Besides the obvious factor of geography, which usually limits a community's reach to those who reside nearby, religious communities may be characterized by the wealth of the congregation, the social status of its worshipers (country club members vs. lower middle class), the size of the community (5,000 or more can attend Sunday services at a megachurch in the U.S., Nigeria, Singapore, and elsewhere), the racial composition, and whether the community offers schooling for children. Churches on college campuses obviously attract a community of worshipers from among the students at that college. Some churches, by reason of geography, may cater almost entirely to immigrants.

Then there is the question of pastoral leadership of the community. Those who serve as God's Regents on earth present as many different leadership and management traits as the general population, and when it comes to religion, congregations may prefer an inspiring speaker first and foremost, over a skilled administrator or budget-conscious pastor. Megachurches often require celebrity pastors who can bring in audiences, not just physically to the church, but to any satellite churches that are part of the community, and through the medium of television and video. There are pastors now who operate strictly through social media, have a million or more followers on the internet, yet have no physical church and often no formal religious training. They earn their money as guest speakers at real churches, through book sales, and through selling of religious trinkets over the internet.

If all this gives you the impression that religion is a business, you are right when talking about megachurches and celebrity pastors, many of whom are multimillionaires who fly on private jets owned by the church, and live in free housing provided by the church. These types of ministries often promote the "Prosperity Gospel," which promises worldly riches, health, and happiness to worshipers, especially if they tithe by giving 10% of their gross income to their church.

Religion as Organized Abuse

The Prosperity Gospel has no sound Biblical support for its teachings, and is thus a form of religious abuse perpetrated on its followers. The same could be said for the many religious cults which form within any religion (not just Christianity). Religious cults are characterized by a strong leader who exercises complete control over the lives of his congregation, is the sole interpreter of religious dogma for the group, often pits congregants against each other, demands that congregants turn over all their wealth and income to the community, isolates members from their outside friends and family, and all too often physically or sexually abuses the members, including children.

Within the U.S. alone, stories appear frequently about religious cults being brought into the public limelight because of a trauma or death inflicted on one of its followers. The abuses of the Prosperity Gospel, and the general greed of televangelists, are on display almost daily on American television. For decades the Catholic Church housed pedophiles and covered up their crimes. Once or twice a year a Charismatic or Pentecostal pastor will die after being bitten by a venomous snake. It is estimated that 1,000 American teenagers every year attempt suicide because they are told they are sick or mentally deficient due to their sexual orientation. Many of these children are kicked out of their homes by their parents and wind up living a life of prostitution and drug abuse on the streets. One can always find situations of abuse or criminal behavior among religious communities.

Yet for every appalling situation, there are hundreds more religious groups which worship God reverently, respect all members, and operate under the leadership of a conscientious, kind, and ethically sound pastor. One could say that religion is a human endeavor which is prone to all the faults found in other organizations. Religion, however, holds itself up to a much higher standard. Its product is belief in and worship of God. It claims for itself ownership of society's moral well-being. It purports that its Regents are endowed with a special personal holiness, because they speak directly to, and on behalf of, God himself.

Religion can ill-afford any stain on its reputation, or faith in the religious denomination or community can easily collapse. The Catholic Church is an excellent example of this reality; it will be many years before its members, and the general public, forgive the Church for the crimes of pedophilia that occurred for decades within what is the largest religious denomination in the world.

There is more to this problem than simply saying all churches are staffed by fallible humans. The problem starts with the fact that the Community function of religion easily slips into an emphasis on tribal identity that closes off the church from the outside world. Newcomers who move to the Southern U.S. often find that the first question they are asked is, "What church do you go to?" This is not a religious question as much as it is a tribal question; the inquirer wishes to know about the cultural identity of the newcomer – where they live, what their income might be, with whom they are likely to be friends, and how their children are being raised.

The second problem is that the Community function of religion is often in conflict with the Regency function. A strong-willed pastor can very quickly transform an outgoing religious community into an inward-looking suspicious community of worshipers. The Regent, in other words, has the power to override the best features of a religious community, and emphasize instead the tribal nature of the congregation, which can appeal to the worst features, including shunning people who fall afoul of the Regent's standards or who no longer profess belief in the group's religious dogma.

It must also be emphasized that no matter how strong a board of overseers might be, the Regent always speaks for God and has the presumption of divine authority that allows him or her to establish whatever dogma or practices they wish. Those who disagree can always leave, is the usual response by the Regent in such circumstances, which is a statement always associated with pastors who are more interested in control than responsible leadership, management, or pastoral care.

A religious group, in other words, is not like a typical for-profit or non-profit enterprise. Once God enters the picture, through whatever cultural filters he must pass in order to be acceptable to the community which worships him, the potential for abuse expands enormously. This is because he who speaks for God cannot be contradicted or countermanded, unless their abuse becomes intolerable or comes to the attention of secular authorities.

The fact that abusive Regents and congregations are a small number among thousands of religious groups does not lesson the gravity of their predations. All of religion is tainted by scandals which occur among a minority of congregations, because the halo of righteousness that is shared by all religions is dependent on the good and decent comportment of all religions. Religion is supposed to be a force for good in the battle against evil; it turns out to be a force for good tainted by those who are religious and who practice evil.

Indeed, it is increasingly becoming evident that it is impossible for religion to avoid an association with scandal because religion inherently promotes selfishness among its children, over altruism. In a 2015 study published by researchers at the University of Chicago, it was discovered that children raised in a religious environment were less altruistic than children raised in a secular environment. 1,170 children from six countries, representing Christian, Muslim, or non-religious families, were tested on their tendencies to share with children they did not know, and their tendencies to punish other children for bad behavior. In the six countries tested (U.S., Canada, China, Jordan, South Africa, and Turkey), in every country the religious children were *less* altruistic than secular children, and *more* willing to punish. Moreover, the older children who were exposed to religious training longer, were even less altruistic and more punishing than the younger children, suggesting that further exposure to religion does more damage to natural human altruistic impulses. This was, in fact, the opposite of what the researchers expected to find, since they were attempting to confirm that religion inculcates moral behavior in children compared to children raised in a secular household. Meanwhile, the parents of religious children were interviewed separately, and consistently believed that their children were more altruistic and forgiving than was the case.

What is shown here establishes yet again that the Community function of religion can frequently promote tribalism, which is defensive and inward-looking by nature. At an early age, religious children learn about sin, good and evil, heaven and hell, and God's horrific punishments for violating his commandments, which are in truth the commandments imposed by and interpreted by the adults in the community. They learn early on about their own immoral and intrinsically sinful nature, which they can never entirely shed as believers in God. They learn that if they follow these commandments, and obey their community leaders and especially their parents, they will ultimately be rewarded, which will not be the case for many children outside of their tribe who do not have the secret to achieving immortality with God that their community enjoys. So incessantly and so diligently is this teaching promoted, that, as Arthur Schopenhauer was the first to note, it is virtually impossible for a child in such an environment to ask, Is it True?

Similarly, nothing in this environment encourages children to think for themselves about morality, and about good and evil. They are allowed to discuss these topics, and tinker with the interpretation of important moral issues around the edges as it relates to their own life, but the fundamental philosophy of good and evil has been thought out for them completely long before, and that teaching, which is often positioned as a Manichean philosophy of a world divided only into good and evil, is inviolate and immutable.

This brings us to an important question: what are good and evil, and how does religion define and use these concepts for its own theological purposes? Since we have concluded our discussion of religion by introducing this important question, it is now time to turn to a discussion of good and evil, and equally important, a discussion of misfortune, and see what we can make of these concepts.

PART THREE

FAITH AND ATHEISM

CHAPTER FIFTEEN

What are Misfortune, Evil, and Good?

*Each of us has a vision of good and of Evil. We have to
encourage people to move towards what they think
is good... Everyone has his own idea of good and Evil
and must choose to follow the good and fight Evil as
he conceives them. That would be enough to make the
world a better place.*

- Pope Francis

Bipolar Beginnings

Discussions of morality and ethics tend to focus on the problem of good and evil. Often these two concepts are described as good *vs.* evil, reflecting the conviction that each is the exact opposite of the other, and that an eternal conflict exists between the two. This conflict, however, is a religious artifact, and its origin can be traced far back in time, to the teachings of Zarathustra, the Persian religious figure who lived around the late 7th century BCE. What little we know of him is to be found in the few surviving texts used by the priests of the time, known as Parsees. These texts refer to two deities, Ormuzd and Ahriman, who fought an eternal, never-ending battle for control of the cosmos. Ormuzd was the god of light, and the font of all goodness in the world, while his enemy Ahriman was the god of darkness and all that is wicked and evil.

Zoroastrianism, as this religion is called, fell into decline after Alexander the Great conquered Persia in 333 BCE, but elements of its dualistic view of the universe found their way into Judaism, and eventually into Christianity. The earliest Christians entertained a great variety of views regarding the cosmos, and the role played by good and evil. One such Christian, referred to as the Holy One, or Mani, arose in the middle of the third century CE in Persia. He created a synthesis of all the religions known to him at that time: Christianity, Gnosticism, Zoroastrianism, ancient Babylonian concepts, and Buddhism. Manicheism, the resulting religion, adopted the idea of an eternal battle between good vs. evil as an explanation for their presence in the world, and Manicheism spread rapidly throughout the eastern parts of the Roman Empire.

Where did God fit into this view of the world, if good and evil existed co-equally in separate realms? Manicheism resolved this by allowing the god of all that is good – the Father of Majesty in Manichean terminology - to play a passive role in the battle of good vs. evil. The Father of Majesty was the equivalent of the Christian Supreme Being, God the Father. The Father of Majesty created a Mother of Life, who in turn created the first man, Adam. It was Adam who donned armor to fight Satan and his forces of evil. Adam periodically appealed for help whenever Satan got the upper hand in the battle, and the Father of Majesty accommodated his requests. The Manichean theology was considerably more complicated than this description, but one important element of this theology was the fact that the Father of Majesty battled an equal in power in Satan. They operated in an upper and a lower realm, respectively, and their battle was never-ending. They were very similar to Ormuzd and Ahriman.

Eventually Mani, having developed a synthesis of all known religions, renounced Christianity except for those elements included in his own religion. He renounced his Babylonian heritage as well, which caused him to be exiled by the Persian king. He wandered for many years around the Near East, establishing communities devoted to Manicheism, returning every so often to Babylon, the capital of the Persian Empire. When Bahram the First became king, he persecuted Manicheism throughout the Persian Empire. Around the year 276, he had Mani crucified. This cruelty, as one might expect, strengthened the resolve of Manichean communities to struggle on, and Manicheism thrived in the Near East for many hundreds of years afterward, long after Christianity became the dominant religion of the Roman Empire.

As to the question of a cosmic battle between the forces of good vs. those of evil, Christianity developed along somewhat different lines. As Jesus began to be seen more and more as a deity, God the Father needed to be better defined, as did that of his messenger to humans, the Spirit (or Holy Spirit). Was the Father superior to Jesus? If he existed eternally, did he precede Jesus? This would make sense if God the Father was also the Creator, who could "give birth" to his son, Jesus. By the time St. Augustine in the 5th century CE was writing his *Confessions*, Christianity had resolved most of these issues. God the Father, Jesus his only-begotten Son, and the Holy Spirit, were one and the same aspects of the Godhead, or Trinity. They were equal to each other, but had different roles. They were each an element of God, and since God was all-Holy, each element of the Trinity was the source of all good in the universe.

As to evil, Christianity eventually rejected Manicheism's belief that good and evil were equal, operating on separate planes. Within Christian theology, evil was represented by the Prince of Darkness, Satan. Satan was a creation of God; at one time one of his angels. He was therefore subordinate to God, and would ultimately succumb to God's power when at the End of Time God would end evil's existence on earth. As in both Zoroastrianism and Manicheism, Christianity discovered a purpose for Satan, as God's opponent in battle, this time over the souls and fate of men. At the end of this battle,

Satan would reign only in his realm, where he would torture for eternity all those souls he had been able to conquer.

Theologians like to describe Zoroastrianism and Manicheism as philosophies which promoted *Dualism,* the concept that the cosmos was divided into two competing realms, destined to fight each other for eternity. Clearly, Christianity did not accept the argument that God and Satan, representing good and evil, were equal players for all time, but Christianity did accept the argument that God and Satan were engaged in a battle for our souls.

In human terms, and in our time scale, as long as this battle continues during our lifetime, we live in a Manichean universe, in which good and evil are equal forces, and God (i.e. Jesus) is fighting Satan on equal terms. This is of critical importance to understanding the mindset of Christians, but also of Westerners, for Western philosophy is steeped in this acceptance of the good vs. evil dichotomy. This concept is so pervasive in our thinking, and so embedded in our culture as Westerners, that we scarcely think about it. Much of our story-telling is based on the struggle between good and evil – it shows up constantly in our books and movies. We may not technically live in a Dualist universe, but Dualism governs our moral and ethical views, because it is so entwined with our definitions of God as all-holy, and his nemesis Satan, or the devil, as all-evil.

This philosophy – let's call it Dualism for practical purposes even though theologians will quibble – is our inheritance from Zoroastrianism, Judaism, Christianity, and ultimately Manicheism. Dualism, however, was not how the rest of the ancient world looked at questions regarding morals and ethics, or good and evil. There was another governing principle that the ancient Greeks, and their successors the ancient Romans, understood and respected, and that was the role that *Fate* played in our world and thus in our lives.

The Greeks identified Fate as a trinity of sorts, called *Moira,* which consisted of three sisters whose role as immortal goddesses was to control the lives of men – their birth, their destiny, and their death. The Greeks were somewhat conflicted about Moira – sometimes the sisters controlled men, but other times the other gods, from Zeus on down, could interfere in men's lives as well. Moira was not superior to Zeus, but was by no means dependent on Zeus.

The presence of the three Fates in human affairs created a sense within each human of *fatalism.* Fatalism was the expression of the reality of our existence, which was that we did not control the two most important parts of our lives – our birth and our death (setting aside our right to suicide), and much of what happened to us in our lives seemed to occur by chance. Whether our destiny was pre-ordained by the Fates was an interesting philosophical point, but relatively unimportant, if we had no way of knowing what our destiny was, and no way of fundamentally altering that destiny. The only way out of the gloomy reality of fatalism was the prospect of appealing to Zeus, Hera, Apollo, Aphrodite, or the other gods to intervene in our timeline in some way, but even these gods could not alter our fundamental destiny. None of us humans, in other words, is immortal.

The Greek philosopher Epicurus, in light of the depressing reality of our existence as described by fatalism, chose to emphasize the need to procure as much pleasure in one's life as would be both possible and virtuous. Epicureanism has often been criticized as a philosophy that allows the indulgence of all sorts of vices, such as lust, gluttony, and the worship of material goods. This is not what Epicurus taught. He stressed the need for moderation in one's appetites, without which it was not possible to act in a virtuous way.

A contemporary of Epicurus, who wrote in the 3rd century BCE, was Zeno, who developed the school of Stoicism. The Stoics looked more closely at Nature, and understood its capriciousness as somehow derived from the will of the gods – of the Fates, more specifically. Man could not control Nature. The best he could do is live his life *in conformance* with Nature, by accepting its random and capricious effects on our lives, but at the same time to cultivate an indifference to these effects, and ultimately an indifference to death. A Stoic lived their life with courage, and with a desire for wisdom, because knowledge of one's circumstances provided a better means of living within conformance of Nature's demands.

The Romans found both these philosophies attractive, but it was Stoicism more than anything which appealed to the Roman imperial mind. As rulers of an Empire, the Romans were intimately involved in military conquest and military governance of the entirety of the Mediterranean world, and militarism was a hard discipline, requiring the opposite of sensual indulgence. Roman leaders, from the Emperor on down, were expected to sacrifice their lives if necessary for their country and their families, and the long list of Roman leaders who committed suicide when Fate dictated no other solution for them (Cato, Marcus Junius Brutus, Marc Antony, Seneca, Nero, for example), is testament to the power of the Stoical mindset.

The history of Western philosophy since the Romans has been one of ignoring, for a very long period of time, the contributions the Greeks and Romans made to ethics and morals. This is because Western philosophy from St. Augustine on to the Enlightenment, was based on Christian religious theology. That meant that the struggle between good vs. evil was the paramount intellectual approach to morals and ethics, which focused on living a life that best achieved salvation through the sacrifice of Jesus Christ, to gain eternal life with God the Father. Morality was interpreted through the filter of the Ten Commandments.

For our purposes, if we are going to understand good and evil, we cannot do so solely through the Ten Commandments or religious thinking about morals and ethics. We must expand our vision to incorporate philosophical thinking which developed in the Enlightenment and beyond into the 19th and 20th centuries. It is not that we must become Epicureans or Stoics, but that we must understand something of the universe in which we live, under a God of our own creation and design.

To begin this discussion, we have to add a third concept to our review of morals and ethics, and that is the concept of misfortune. We can call this Nature, or natural law, or even fate, but we must begin by investigating the capriciousness of Nature.

What is Misfortune?

Misfortune is a consequence of the natural workings of the universe. It is not the only consequence. Beneficence, the opposite of misfortune, certainly occurs as well. One can think of misfortune, as it applies to humans, as bad things happening to people unexpectedly.

We are all familiar with misfortune and the varying degrees of unhappiness it can cause. On the mild side, misfortune can be a nuisance: the car won't start in the morning, making us late for work; the fruit we bought in the supermarket spoiled on us in less than a day; bad weather causes our flight to be delayed for hours. In cases such as these we are inconvenienced, or our daily routine is disrupted, or we cannot meet the expectations of others who are depending on us. We don't perceive of nuisances as misfortune, but they are because of the unexpected nature of the inconvenience.

Misfortunes that are more than just a nuisance often have a long-lasting character to them. If we suffer a significant financial loss in our retirement investments, it may take years to make up for the monetary setback. Chronic damage over the years to our body is difficult to reverse, even if we try surgery, which is not always successful. In both these cases, the fault for these problems may be our own. We may have made manifestly injudicious investment decisions, putting our money in highly risky, speculative stocks. It might be that our body deteriorated because we worked in a strenuous and dangerous job prone to injuries, and we were more careless than others around us. Still, it is difficult for most people to blame themselves entirely if at all.

We like so much to assign blame and responsibility for misfortune that it is something of a second nature to us. In moments of candor we might say to a friend: "I made some really stupid investments." More typically, we will opt for the less condemnatory "I lost some money in the stock market." Everybody knows the stock market goes down as well as up, and that its workings are mysterious, unpredictable, and capricious. Merely telling people we lost some money in the stock market lays the blame on the capriciousness of the financial markets, and away from ourselves.

This stock market example is a particularly useful instance of the effect of randomness in our lives. The stock market is a human endeavor that is demonstrably and acceptably defined by its random behavior. Financial experts and analysts spend their careers using random variables from statistics as a means of explaining and predicting the stock market. They might say, for example, "based on the past twenty years of statistics on this stock, it has a sixty-six percent chance of moving up or down by $5 this next month." Some analysts might tell you only the happy news, that the stock has a reasonable chance of going up \$5, but a proper assessment of the situation has to reflect the possibility of gain and loss, since that is how stocks in a freely-trading market work.

What is interesting about this example is how universal the circumstance is, but how rarely we apply it to our lives. Meteorologists have well-crafted and reasonably accurate models of the likelihood of powerful hurricanes or cyclones forming in the Atlantic or Pacific Oceans. Some years the models might predict a mild hurricane season in the Caribbean; other years a violent series of storms is expected. The year Hurricane Katrina hit New Orleans was one in which more violent hurricanes were to be expected. As the hurricane developed, the models were able to predict the likely point of landfall, but even a few days before landfall occurred, no meteorologist could say for certain that Katrina was going to strike precisely where it did, causing the damage that ensued. Scientists were working, after all, with probabilities based on many past experiences, but these probabilities never guarantee a particular outcome in the future.

Over 1,000 people died as a result of Hurricane Katrina, with many more injured, amid massive property damage that changed the structure and face of the city. Misfortune is hardly an adequate description of what happened to New Orleans. This natural catastrophe continues to have repercussions for the victims, not least for those who lost family or friends in the floodwaters that inundated most of the city. What does one say to those left behind to deal with their grief as a result of this hurricane?

The cold reality of Hurricane Katrina is that it was a natural disaster. It was, to some degree, predictable. Over a very long span of time, involving many centuries, a storm like Katrina has hit the New Orleans area and will strike again. We know this from a statistical review of weather patterns in the Caribbean Sea, and the frequency with which Category 5 storms appear among these weather patterns. We know as well that expansion of human settlements along the coasts of hurricane-prone areas is going to increase the risk of loss of life and property. We know further that climate change may be worsening these storms and altering the statistics. Perhaps now it might be the case that a hurricane like Katrina will hit New Orleans every 100 years, or even every 50 years. It is easily possible to say that for each such storm, 1,000 people or more might die.

So do you say to the victims' surviving family and friends? – "too bad; sorry that you were a statistic, a random victim of this storm." No one talks this way, partly because someone suffering deep grief needs personal and genuine solace to comfort them, not callous indifference to their pain. But to repeat – the cold reality was that some people in New Orleans were terribly unlucky during Hurricane Katrina. They might have had a propensity to be poor, since the rich learned a long time ago to occupy the high ground in New Orleans, away from the places prone to flooding. Nonetheless, even among those living in low-lying areas, it seemed quite random which individual was destined to die.

We humans have an almost insuperable difficulty accepting the role of randomness in our lives, especially when it comes to any disasters that befall us. In the face of natural disasters, severe accidents, unexpected terminal illnesses, and the death of someone close to us – we are desperate for a reason. The correct response would be for us to blame the universe – the illimitable and utterly impersonal cosmos that deals out misfortune regardless of whom or what is in the path of destruction. This, however, we cannot accept. All of our instinct is to seek out cause and effect. There must have been a reason for this injustice dealt to us. Moreover, we have a need to anthropomorphize such circumstances. If we cannot identify a particular individual or group who visited this pain upon us, we must put a human face and a human personality to whatever caused this. If it is indeed the cosmos which is to blame, we need to humanize the cosmos in order to understand what happened, and perhaps avoid it in the future.

As we've seen, the Greeks solved this problem by inventing the three Fates. Christianity may not have three Fates, but it has three gods in one. The triune God knows when each life will end, since God is presumed to know all things. This God therefore knows all misfortunes which may befall a man. In a sense, one can say that the misfortunes are pre-ordained, since God knows of them in advance and they will certainly occur. The Fates, however, were passive observers of what happened to mankind. The God that Christianity creates is not. He is our loving Father. He wishes that we will join him in heaven for all eternity. He has already intervened in human history through the Incarnation of his son. This gives him an active interest, or investment if you will, in the life we live, the decisions we make, and the nature of those decisions. Those who decide wisely, those who live morally, will be rewarded in heaven.

Why, then, does he preordain misfortunes to occur? What is the purpose of any misfortune? There are no answers to these questions, no matter how many people write books attempting to resolve the enigma of misfortunes. A loving Father would not deliberately torment his children. He would not willingly inflict accidents or sickness on them. He would not condemn babies to death due to birth defects, poverty, or warfare. Nor would he do any of these things as "tests," to see whether the individual can suffer enough to be worthy of admission to heaven.

Christianity takes from the story of Adam and Eve a lesson that explains human suffering and misfortunes. As Adam and Eve violated God's sole commandment not to eat from the Tree of Knowledge of Good and Evil, Christianity teaches that all of mankind must pay a price. In some denominations of Christianity, the guilt is collective and is called Original Sin, depriving each human of the right to heaven unless they work hard for it during their lifetime. In other denominations Original Sin is less stringent – it means only that each human is born with a propensity to temptation and sin, and must learn to live a righteous life. Again, why would a loving Father visit the sins of Adam and Eve on all of us who follow? This is the enigma – the conflict between the Christian belief in God, as a loving Father who never would harm us, against the fact that we will sometimes be random victims of the universe, which inflicts unpredictable, bad things on us all, until the final calamity of death, which is one of the presumed consequences of mankind's expulsion from the Garden of Eden.

There is only one way to resolve this problem. We must recognize that the loving God we postulate to exist outside of ourselves does not do so. He exists only in our mind, and even then only as Illusory God – an illusion. The conflict that we think exists between a loving God the Father and the misfortunes we see all around us, exists only in our imagination. Outside of ourselves, there is no conflict. There is, instead, something that is exceedingly difficult to accept, and that is the randomness of misfortune, which can strike us down, or those we love, at any moment. We can never truly anticipate or prepare for such misfortune. We can only learn to expect it and accept it, including the misfortune of our death.

To do otherwise is to create a never-ceasing argument with yourself. You can never win such an argument, because you are the one conflicted, not the universe. Nor is it of any avail to try to explain and understand the universe further. No matter how many scientific discoveries we make, we cannot change the nature of the universe. The universe is what it is, and we are but a minute part of it. We cannot fathom why some people die in a plane crash and other people walk away. Bad things happen to all people, and the worst thing of all – death – happens to everyone. That is the way things are, and our life is better spent accepting this reality than constructing fruitless arguments in our mind, trying to invent reasons for the unexplainable.

Exactly how we go about creating such a life is up to us. We discussed earlier two ancient paths that philosophers for the past few hundred years have investigated: Epicureanism and Stoicism. Both of these concepts have been refined and modernized, and expanded upon, to include for example the idea of materialism, which is the philosophical belief that the universe is composed of minute building blocks of matter, which constitute the only reality of the universe. This leaves out the possibility for a supernatural or guiding force, and it also means that the basic materials of the universe, (atoms to the ancient Greeks and subatomic particles to modern man) interact in a random fashion. At a higher, more complex conglomeration of atoms – you and me for instance – this randomness as it affects us can constitute misfortune, or beneficence.

Misfortune, therefore, can happen to any of us, at any time, and with any degree of harm. That is the nature of the universe and there is nothing we can do about it except act with reasonable care and caution in our lives. There are circumstances, however, when bad things can happen to us that are caused by people, and not by the universe. Harm caused us by the universe is extremely difficult for us to accept because it is so random and unexpected. Harm caused us by other people is exceedingly difficult for us to understand, but fortunately we do not have to accept it. We can seek out punishment for the perpetrator to prevent further harm to others. When we try to understand harm caused by other people, we seek out a quality that characterizes such people, and we call that quality *evil*, for harm that is particularly grievous and deliberate.

What is Evil?

Evil is the deliberate infliction of grievous misfortune by one person on another person or living thing. Evil is therefore the consequence not of Nature *per se*, but of human nature, wherein a human steps in and circumvents nature, by bringing misfortune on to another person. To be evil, a human action has to be deliberate, with the intention behind it of causing harm. Someone who accidentally causes you to trip and fall down is not perpetrating evil. It requires a deliberate intention to harm that qualifies an act as evil.

The second consideration is the grievous nature of the act. Evil is usually considered most serious if bodily injury or death is a consequence of someone's deliberate actions to harm. Financial con artists may destroy the life savings of hundreds if not thousands of investors, but most people would rank such action as low on a scale of evil activity. The victims at least have their life intact, and have an opportunity to rebuild their financial condition. Not so with the victims of mass murderers, genocidal dictators, serial killers, and sadistic torturers. The greater the scale of death and harm that is dealt by the perpetrator, the more we are inclined to identify the criminal as evil.

Why people perform evil acts is never easy to determine. In ancient times, it was widely believed that people could be possessed by a devil which caused them to do deliberate harm to others. Jesus certainly held to this opinion and provided healings or exorcisms for such people, and even for animals (the herd of Gadarene swine, for example). In modern times, we have a great deal of medical and related research that indicates people may have inherited genetic mutations that compel them to psychopathic violence. This research has brought into question the idea that we all have free will, and can make a choice before performing an act of evil. Some people may have very little choice, and there are cases of criminals who have admitted to rapes or multiple murders, who plead for medical castration or even the death penalty to prevent them from performing further evil acts. While there are 2,000 years of history between these two approaches, they have one thing in common. They both find the source of evil in the mental imbalances or disturbances of those who perpetrate evil.

What of those who perform great evil on a mass scale – men like Stalin, Hitler, Mao Tsetung, Pol Pot? There is something other-worldly about such individuals, because their crimes are unfathomable to the rest of us. They do not appear to be part of the human race, and even assuming they had psychological reasons for doing what they did, we have a hard time forgiving them or justifying a lenient punishment. Quite to the contrary – they are branded by religious believers as Satanic, as some incarnation of the devil.

We will discuss later the religious belief in divine punishment for evil, including the condemnation of evil people to eternal hell. For now it should be said that certain notoriously vile individuals like Hitler or Mao, escaped trial and punishment for their crimes, thereby challenging a believer's faith in God. Christians ask how God could allow such men to die peacefully in their bed, or at their own hands rather than that of an executioner. The only answer to this question, if we are religious believers in God, is to assert that God has in store for these people eternal punishment in hell. Many people derive some comfort, when thinking about Adolf Hitler, in assuming that at this very moment and for evermore he is suffering the hideous fires of hell.

What if we lived in a world where few if any people believed in God, hell, or eternal punishment? In such a world, no one would want Hitler or any other mass murderer to "get away with it." Governments or individuals might be more likely to stop someone like Hitler before he died, and history might then be different. A belief in divine punishment can sometimes be an excuse to do nothing when faced with evil.

When the evil is great enough, moreover, a believer's confidence in God becomes shaken, and they wonder again how he could allow so much evil to exist in one person. Because the horror of what has been done is so great, believers are obliged to think that there is some other force in the universe completely different from God. They are obliged to think of the devil, and ascribe to men like Hitler a Satanic quality that matches the horror of their crimes.

As God and heaven are one and the same and inseparable, so the devil and hell are inseparable. What is the nature of hell, and what role does the devil play?

What is Hell?

Hell is the fear of death magnified to monstrous proportions, the purpose of which is to enhance the power of religious Regents over their followers. Death is the greatest of all anxieties we face, and so appalling that we suppress thoughts of death as much as possible in order to carry on with our life. Death is not only the extinction of all we hold valuable, it is an event shrouded in the greatest uncertainty. What, if anything, will follow our death?

We are prone to answer that question with "not-death." We like to believe we can escape our mortality by viewing death as nothing more than a stepping stone to another existence, or some different form of existence. In fact, this is our default hypothesis, derived from our earliest, and now unconscious memories, of a life in heaven when we were infants. It takes some degree of courage to reject this hypothesis, and to accept that death is the sheer obliteration of our physical being and of our consciousness, with no afterlife to follow.

Since most people prefer to imagine the existence of an afterlife, and to project on to that afterlife a return to the heavenly existence of our infant years, would it be sensible to believe that all humans will be equally awarded with an afterlife in heaven? Absolutely not, when it is understood that believers in God would prefer not to spend their afterlife with certain other people. Some people have led a life of evil, imposing cruelties on others, and they are not only not deserving of a reward of an afterlife, they should not be allowed to pollute the afterlife of everyone else with their presence in heaven. In other words, the afterlife, existing as it does for all eternity, has to have some sort of admissions policy, based on whether an applicant conducted their life in a manner consistent with good behavior, or evil behavior.

This is the reality of any belief in the afterlife; it has to take into account how any particular person lived their entire life. If you wish to believe in God, and you wish to believe that you can cheat death by passing into the afterlife after your own demise, there are simply going to have to be some rules regarding admittance, to avoid both the unfairness and the unpleasantness of evil people being permitted to enjoy the afterlife with you. This is yet another reason for the invention of religion; someone has to be the arbiter of the nature of the afterlife that each person will ultimately be allowed to have, and religion provides just such a person: the Regent, speaking on behalf of God.

It follows logically that the afterlife must be divided into two realities, and two physical locations: one for the just (which presumably means believers who led a good life), and one for those who are evil. We can define heaven as the return to the blissful realities of our infant and early childhood years. Evil people must by definition be denied this bliss, just as they must be denied oblivion, which would be a reward of a sort for their evil life, since they would be granted eternal sleep, and everlasting unconsciousness. Indeed, and again all of this flows logically from anyone's basic belief in an afterlife, those who are evil must be consigned to a place of punishment, which will last for all eternity, since the afterlife for the rest of us lasts for all eternity.

What shall the punishment be for the unjust, for those who are evil, for those with whom a believer in God would not wish to share eternity? The punishment is to condemn them to hell, or Gehenna/Sheol (in Judaism), Tartarus (in Greek mythology), Jahannam (in Islam), or Naraka (in Buddhism). The punishment that is meted out in these places is entirely up to the consigner – to that person who defines the nature of hell, and the type of person condemned to spend eternity suffering in hell.

The ancient Greeks distinguished between two places in the Underworld. The upper location was Hades, ruled over by a god with the same name. Hades was the abode of the spirits of the dead, none of whom was ever allowed to return to the land of the living. It was a doleful place, but no tortures were inflicted on the dead. These were reserved for those condemned to Tartarus, located below Hades, and reserved originally for the Titans who rebelled against Zeus. This was one of the earliest examples of blasphemy – rebellion against God – being punished by damnation. The punishments were on the order of frustration as well as physical pain, and as eternal punishments go, few cultures have been more imaginative than the Greeks, and less interested in pure torture as punishment. King Tantalus, for example, who served the flesh of his slain son to the gods in a feast, was forced to stand forever in a pool of water, which would recede from his mouth if he attempted to drink from it, while a tree with fruit would withdraw from his reach if he was hungry.

In Jewish scripture, Sheol was the abode of the dead, very much like Hades, while Gehenna was a place of judgment for the wicked. The references in the Old Testament are limited, but describe Gehenna as a "pool of fire." This is the earliest suggestion in the Judeo-Christian-Muslim tradition that hell is a place of eternal torment by fire, which most people would agree is the worst pain imaginable. Jesus Christ makes nearly a dozen statements in the Gospels regarding Gehenna, the permanent home for the wicked, such as the Pharisees who opposed him. When you combine the Jewish concept of Gehenna as a pit of fire, with Jesus' many references to this place of punishment, you arrive at the Christian concept of hell. Hell has been a dominant feature of Christianity, and Islam which followed, ever since.

The terrors of hell are an easy sell. It was quickly discovered by the early Christians that hell was a useful tool for encouraging conversion to Christianity, as well as an ongoing commitment to the faith. This is because hell plays on three horrifying fears: that of death, that of pain, and that of a combination of the two for all eternity. As we mentioned above, any acceptance of the idea of an afterlife automatically entails the acceptance of two afterlifes, one for the just, and one for the wicked. The idea of an eternity of pain, especially something as unbearable as pain from fire, is the ultimate human horror.

God's Regents on earth learned how to exploit the fear of hell, even to the point of threatening children with hell. Here are some notable examples, starting with one of the earliest Christian writers, Tertullian, who delighted in a vision of Christians triumphant over their defeated Roman and pagan masters:

> **What a spectacle. . .when the world. . .and its many products, shall be consumed in one great flame! How vast a spectacle then bursts upon the eye! What there excites my admiration? What my derision? Which sight gives me joy? As I see. . .illustrious monarchs. . . groaning in the lowest darkness, Philosophers. . .as fire consumes them! Poets trembling before the judgment-seat of. . .Christ! I shall hear the tragedians, louder-voiced in their own calamity; view play-actors. . .in the dissolving flame; behold wrestlers, not in their gymnasia, but tossing in the fiery billows. . .What inquisitor or priest in his munificence will bestow on you the favor of seeing and exulting in such things as these? Yet even now we in a measure have them by faith in the picturings of imagination.**
>
> **-Tertullian**

Later, when Christianity was triumphant over the secular world, preachers had no one to threaten but the unbelievers, or the unrepentant sinner:

> **The world will probably be converted into a great lake or liquid globe of fire, in which the wicked shall be overwhelmed, which will always be in tempest, in which they shall be tossed to and fro, having no rest day and night, vast waves and billows of fire continually rolling over their heads, of which they shall forever be full of a quick sense within and without; their heads, their eyes, their tongues, their hands, their feet, their loins and their vitals, shall forever be full of a flowing, melting fire,**

fierce enough to melt the very rocks and elements; and also, they shall eternally be full of the most quick and lively sense to feel the torments; not for one minute, not for one day, not for one age, not for two ages, not for a hundred ages, nor for ten thousand millions of ages, one after another, but forever and ever, without any end at all, and never to be delivered.

-Jonathan Edwards, whose sermons, according to local newspapers, caused some parishioners to flee the church building and commit suicide.

Some pastors took a special delight in tormenting children:

Little child, if you go to hell there will be a devil at your side to strike you. He will go on striking you every minute for ever and ever without stopping. The first stroke will make your body as bad as the body of Job, covered, from head to foot, with sores and ulcers. The second stroke will make your body twice as bad as the body of Job. The third stroke will make your body three times as bad as the body of Job. The fourth stroke will make your body four times as bad as the body of Job...The fifth dungeon is the red hot oven. The little child is in the red hot oven. Hear how it screams to come out; see how it turns and twists itself about in the fire. It beats its head against the roof of the oven. It stamps its little feet on the floor

- The Rev. J. Furniss

It was especially important for these observers to emphasize that the damned maintained their consciousness and sensibilities through every moment of torment, because it wasn't just the body being tortured, it was the soul. This was an observation made by John Calvin, John Bunyan, and Ignatius Loyola. It was also noted that the righteous would thoroughly enjoy the "spectacle" of watching the wicked suffer in hell. This was especially emphasized for those who questioned whether God in his mercy would ever allow a parent to be separated by one of their children condemned to hell. Quite to the contrary!

Once [a soul] is condemned by God, then God's friends agree in God's judgment and condemnation. For all eternity they will not have a kind thought for this wretch. Rather they will be satisfied to see him in the flames as a victim of God's justice. ("The just shall rejoice when he shall see the revenge . . ." Psalm 57:11) They will abhor him. A mother will look from paradise upon her own condemned son without being moved, as though she had never known him.

- St. Anthony Mary Claret

Frightening images of hell have done their work for over 2,000 years of Christian history. They are no longer as necessary today as five hundred or more years ago. Most modern audiences know exactly what hell is like, having seen representations in the movies. In order to scare someone, it is merely enough for a preacher to mention hell, with a special emphasis on the fact that the listener must make a binary choice: to receive Jesus in their heart and be rewarded with heaven, or to reject him and be condemned to hell.

Hell is just as real as you are sitting wherever you are at, right now. You have a destination. You have an eternal home. You are marked to go somewhere, Heaven or Hell... You are one breath away from eternity... Time is running out... Some of you are already heading to Hell. I have family and friends who are heading straight to Hell... We are all sinners – separated from a righteous God. God is wrathful. He is a God of vengeance. Yet he thinks, how terrible it would be for you to go to Hell. So he sent his only Son to pay the price for your sins. God has done everything he could to save us. He said, "Come with me." [But it is] you who must decide, and who must take action. *You must approach God with the faith of a child* {italics mine}. The only way is to confess your sins. Give him your life, so he can [in return] give you eternal life [with him in Heaven].

-YouTube Evangelist Clayton Jennings, 2015

In fact, some believers do not have to hear a sermon from a pastor or imam to be reminded of the existence and threat of hell. Here is the American boxer Muhammad Ali, describing his personal reminder of the hellfire of Jahannam:

I don't smoke but I keep a match box in my pocket. When my heart slips towards sin, I burn the matchstick and heat my palm with it, then say to myself, 'Ali you can't even bear this heat, how would you bear the unbearable heat of hellfire?'

-Muhammad Ali

Modern theological interpretations of hell have moved away from the traditional view of fire and brimstone, as a reaction against the blatant terrorizing of the believer and non-believer that has been such a common feature of Christian (and Islamic) teaching. Intellectuals in particular tend to believe the punishments of hell are not physical, but mental and emotional. In that sense, they are just as cruel, but more sophisticated. Here are some examples:

The mind is its own place, and in itself can make a heaven of hell, a hell of heaven.

– John Milton

What is Hell? I maintain it is the suffering of being unable to love.

- Fyodor Dostoevsky

So this is hell. I'd never have believed it. You remember all we were told about the torture chambers, the fire and brimstone, the "burning marl." Old wives' tales! There's no need for redhot pokers. Hell is—other people!

- Jean-Paul Sartre

If hell is such a fearsome yet imaginary place, why is it so believable? The answer is that hell builds on something which is already extremely real to us – death. Death is the only aspect of life that is absolutely certain. That certainty serves as a base which easily anchors on top of it anything our imagination wishes to impose. Heaven is easy to impose upon our imagination regarding death, acting as an antidote to its terror, because we have already experienced heaven. We take the certainty of death, and we add an imaginary afterlife based on something that once was real in our life, and we create a heaven that is not only easy to accept without question, but something we can readily desire. Similarly, if we add to death an imaginary afterlife of pain – and all of us have experienced some degree of pain in our life – it is easy for us to accept the claim that hell exists. We can accept the existence of hell without question, and learn to dread hell more than we do death, particularly if the images presented to us are of unrelenting agonies of the most intense kind imaginable.

Religion knows this, and religion exploits the fear of hell with no compunction, precisely because the Regents who preach on and on about the terrors of hell believe in it completely. They approach the subject with absolute conviction, and as we have seen in the quotes above, some approach the subject with a delight in the sense of superiority it gives them that they are among the elect who will avoid hell. Hell becomes religion's most powerful tool for drawing in adherents, especially among children, and for enforcing obedience to the Regents who have temporal power over their congregation.

It seems strange that we are so willing to make a trade-off regarding death. Our belief in God allows us to circumvent our fear of death by imagining that we will cheat death's ultimate consequence – oblivion. But we then replace that existential anxiety with a whole other obsessive worry, and that is our concern about going to hell when we die. The reason we accept trading off one existential dread with another is that the first fear is absolutely real – death will occur, and we cannot avoid it. The second dread is supposedly within our control – obey the rules of God's Regents, and live as blameless a life as possible, and we will cheat death, we will cheat hell, and best of all, we will be rewarded with heaven. Even though the second dread is entirely imaginary and has no basis in reality, the scales seem to weigh more heavily in belief in an afterlife, hell, heaven, and the prospect of achieving eternal bliss in heaven if we are obedient to religion's teachings and demands. So strong is the dread of death, and our willingness therefore to believe in hell, that is a very rare individual who does not even momentarily flinch in fright when confronted with a preacher, or even an acquaintance, who says they are damned to hell if they do not believe in God. Even atheists will admit they are not immune to this fear, though it is temporary. The fear, after all, stimulates our Amygdala, and strikes at our emotional core; the antidotes of logic and reason have to be deliberately summoned within the mind in order to overcome the fear.

Now there have always been some people – Voltaire comes to mind - who were immune to threats of hell, and scoffed at the idea as preposterous. Such people also ridicule religion for selling such a story, and among those who can be most upset at religion for its emphasis on and depictions of hell, are parents of young children. It is very disturbing for any parent, or any guardian, to see a young child traumatized by religious teaching which centers around the graphic tortures of children in hell. For this reason, religion has in the past century or two reduced its intense focus on hell, and has adopted a more ameliorative depiction of hell, along the lines of a place where one experiences the mental or emotional pain of eternal separation from the goodness of God.

Why though, would religion put *any* emphasis on hell? The answer has to do with religion's insecurities about its own theology. Christianity is a perfect example of this problem. Christianity, more than most religions with the exception of Mormonism and Scientology, offers an elaborate mythology operating as theology. Setting aside the Old Testament, and starting with the Annunciation to the Virgin, a Christian is expected to believe in a Virgin birth, the birth of Jesus in a lowly manger, a celestial beacon over Bethlehem which attracts three Magi from across the desert, the slaughter of the innocents by Herod, the flight to Egypt by the Holy Family, Jesus coming of age as a carpenter's son in Nazareth, Jesus' baptism with the voice of God announcing his divine birthright, Jesus' three year mission of miracles (including raising one or more people from the dead), Jesus' prediction of his death and resurrection, Jesus' resurrection following crucifixion, Jesus' visitation to hell, Jesus' return to earth and to his disciples, and Jesus' visible ascension into Heaven.

None of this makes sense on a factual basis, in terms of any of this happening in reality. These myths do make sense if compared to very similar pagan myths of their time, and they make sense as rebuttals to charges from Romans and the pagan world that Jesus Christ was raised only by his mother since his father left him at early age, and Jesus was too poor to be a king. He was said to be nothing but a magician, and his body was stolen by his disciples to make it look like he was resurrected.

What ultimately made this mythology so acceptable to so many people, was St. Paul's providing an intense emotional appeal to the myth, by defining Jesus' death and resurrection as the ultimate blood sacrifice by God, for the benefit of his children, meaning each of us. St. Paul makes it imperative that we accept the mythology as fact, because if we do not place our faith in Jesus and his resurrection, we will certainly die in sin and go to hell.

As we are predisposed to believe in hell, we are predisposed to gloss over the nature of the life story of Jesus Christ, and merely accept it all as historical fact, because that is premised as our only way into heaven.

The Personification of Evil

It takes but a moment's thought to realize that evil in the Abrahamic religions has been anthropomorphized and externalized. Evil is personified in the form of Satan, who goes by quite a number of imaginative names assembled over the centuries: Satan, the Devil, the Prince of Darkness, the Evil One, the Great Deceiver, the Dragon, the Antichrist, the Enemy of God, Lucifer – and most people's favorite, just for the poetic juiciness of the name – Beelzebub!

The concept of Satan does not fit within our model for God-belief. There is an Anthropic God – anthropomorphized and externalized God, but there is no Anthropic Satan. However horrific a child's early years might have been at the hands of cruel, absent, or self-centered caregivers, no child converts that experience into a God-like figure called Satan. Instead, these experiences are converted into a real God – Behavioral God exhibiting bad behaviors.

No child experiences in real life a Satanic figure and lives to tell the story. They would not have survived such evil to live to adulthood. The manifestations of God from infancy to childhood, after all, but especially Template God, are representations of *survival*, not death.

So what or who is Satan or similar personifications of evil? All such figures aren't even illusions, because they carry no element of wish-fulfilment. Instead, they are *fictions*, just like characters in a novel, or depictions in movies or cartoons. Satan is, in that respect, one of the oldest fictional characters in literature of any kind, of the order of the monster Grendel, but predating that figure from *Beowulf* by over 1,000 years.

There is one big difference between Satan and most fictional characters. Even though Satan is not some equivalent form of Anthropic God, Satan has almost from inception been anthropomorphized and externalized. He has been given the body and intellect of a man, with animal appendages such as cleft hooves, horns, and a tail. He has been externalized as well, with many believers in all three of the Abrahamic religions interpreting him to be a personage existing in the real world in terms of actual space and time.

There is also one other quality to Satan that is important to the character, and that is the set of supernatural powers he is assumed to have. Satan can travel about the world in physical and invisible form, taking the shape of animals if he desires, and offering unimaginable riches to select targets if this is what it takes to snatch someone into his domain. Satan's powers are entirely equivalent to God's when he is tempting a human into evil. This has to do, in particular, with Christianity's adoption of what we can call cosmic temporal Dualism – the idea that the forces of good (God) and evil (Satan) battle each other for human souls on an even playing field. This battle, however, is only a temporary situation which will come to completion at the End of Time, when God will establish his dominion over Satan, his abode in hell, and his legion of fallen angels, demons, imps, and other supernatural creatures of hell.

The concept of cosmic temporal Dualism implies that in today's world, the battle between God and Satan is evenly matched, the outcome of which for any particular human depends on a) the persuasive power of the two supernatural beings, combined with b) the ability of the human in question to avoid temptation and choose a path of goodness out of their own free will. Those who fail to make that choice will ultimately fall into Satan's realm and will remain there are the End of Time, for all eternity.

The stakes for mankind are therefore of supreme magnitude. In this cosmic theology, the stakes could not be higher – personal salvation and eternal bliss in heaven, or personal damnation and eternal agony in hell. In our modern neuroscientific terms, this is a battle over an individual's Thalamus (the source of altruistic impulses) vs. an individual's Amygdala (the source of fear and flight to safety). The battle is unfair. Fear for one's safety will almost always trump an altruistic impulse to help others. Because of this, the average person is likely to listen more closely to their fears, and thus decide that a priority in their lives must be to avoid evil at all costs, because the consequence of eternal damnation is too frightful to contemplate.

"This is all to the good," you might say. "People are inclined to tremble at the power of Satan, and thus avoid acts of evil." Not quite. For anyone who accepts the cosmic temporal Dualism inherent in Christianity, fear of arrest, trial and punishment under civil authorities for acts of evil is just as strong a motivating factor for the avoidance of evil as fear of Satan. There is no need for civil society to rely on religious fears to help obtain compliance with the law.

The fear of Satan is an unnecessary emotional burden placed on religious believers that is not justified on secular grounds, and is not even justified on religious grounds. This is because the fear of Satan is not met with an equal, but opposite fervor for doing good deeds. Believers do not associate performing acts of altruism as something which must be done anytime they *avoid* the temptation to do evil. Instead, and this practice is especially prevalent in Catholicism, believers must perform acts of altruism *only after* they perform an act of evil, as a matter of penance for the crime of sin.

This may seem like we are focusing on theological minutia here, but the distinctions matter, because when Satan as the personification of evil is introduced into a believer's life, fear predominates as a consequence. Satan is taken as real, an invisible creature working in the cosmic realm who has the power to infiltrate one's mind, offer an infinite variety of temptations leading to sin, and ultimately drag a believer's soul down to hell and eternal torment. In more paranoid forms of religion, especially with cults, Satan can be seen to operate invisibly everywhere within the world. The believer must watch where they go, who they meet, what entertainment they see, and the books they read, in order to avoid giving Satan a knothole that would allow him to crawl into their lives. Such paranoia can lead to a life of paralyzing fear.

Satan is also used as a ready-made excuse for bad behavior, as in "the devil made me do it." Satan becomes inextricably mixed up in the mind of the religious person with that individual's worthless, sinful self. To the extent that Satan is anthropomorphized and externalized, he is not the personification of the enemy of God, as much as he is the personification of each individual's sense of worthlessness. He is the embodiment of a believer's morally corrupt nature.

When we discuss more fully the consequences of sin, we will see why the believer in God pays such a high price for their acceptance of cosmic temporal Dualism – the fight of good and evil for human souls.

The Quality of Goodness

The word "good" is one of the more versatile in the English languages. Here are some of the many different ways we use the word:

Pleasing: "That was a good dinner."

Skillful: "She was a very good swimmer."

Beneficial: "As a politician, he always worked for the public good."

A Pleasantry: "Good morning!"

Well-behaved: "You're a good dog!"

Positivity: "A good attitude will take you far."

Enjoyment: "Have a good time!"

Health: "Her good eyesight allowed her to spot things far off into the distance."

"Good" is a catch-all word for so many things in our life, that we tend to slap it on any experience or behavior that we find desirable. Good in the various manifestations listed above is certainly a word appropriate for describing God, but used in any of these ways, good is too limiting – it is altogether too human. For the God of our imagination, we seek a definition of good that is much more powerful, and capable of capturing God's infinite and timeless nature.

Suppose we were able to summon up our memories of God from our infant and early childhood years, if but for a moment, to understand better the unique qualities of goodness that God displayed. We know at least five of the qualities most associated with God. We've referred to these before in this book: God is Omnipotent (all-powerful), Omniscient (All-Wise, or All-Knowing), Omnipresent (always present), Omnibenevolent (God is Good All the Time), and the Creator of the universe (he who made us). Of these, the Omnibenevolent God provides us with a definition of perfect goodness, starting with the fact that God would have to be all-good to have wished to create us, provide for our every need, and flood our innermost being with his pure Love.

What are some of the other qualities associated with God that imply goodness? God is flawless. He is without error. He is without sin. Why we do we assign such attributes to God? They all seem to be related to God as the source of love, but they imply a quality that is related to, but somewhat different from love.

What they each imply is the quality of perfection. God is perfection, is the implication. Well, now we are getting somewhere, because this definition of God fits right in with our other primary definition of God: **God is our idea of who the perfect human being would be like.** Flawlessness is a quality related to perfection, so clearly God would be flawless, just as he would be without error. And as to sin, this is a very human quality that we knowingly assign to ourselves – to humans in general – because we are well aware of our flaws, and our potential to hurt other people. We define such actions as sinful, and so God must be without sin if he is also an idealized human being.

Now that we have introduced this topic of God as without error or sin, we can expand these ideas into concepts that draw us closer to our interpretation of God as seen through religion. These concepts define God as all-holy, and as sanctified, or sacred in his perfection. While religion prefers to use words like these to describe God, these definitions pre-date Christianity, and are found in its Jewish origins. They are among the earliest and most ancient writings about God, as for example in the Psalms:

> **Hear my cry for mercy as I call to you for help, as I lift up my hands toward your Most Holy Place.**
>
> -Psalm 28:2
>
> **Who may ascend the mountain of the Lord? Who may stand in his holy place?**
>
> -Psalm 24:3
>
> **The Lord is in his holy temple; the Lord is on his heavenly throne. He observes everyone on earth; his eyes examine them.**
>
> -Psalm 11:4, also Habakkuk 2:20

The Old Testament is replete with references like these, where God is the holy one, and his abode, which is heaven, is not simply a holy place, it is the most holy place, befitting any being who is perfect in all aspects. Where did these ancient writers come up with these ideas? They were expressing their ideas about what the most perfect human being must be like. Just like you and me, their references to the most perfect human being were derived from having met just such a person, in their earliest years. They would have drawn upon their unconscious impressions of that experience, and not their direct memories of that experience, because the memories were of course no longer available to them on the surface, in their conscious mind. They were available, in a sense, as impulses of which they were unaware.

Let us look at one additional, and important quality of goodness that God represents. This is the quality of healing; of restoration.

> **Heal me, Lord, and I will be healed; save me and I will be saved, for you are the one I praise.**
>
> -Jeremiah 17:14

References to God's restorative power are also common in the Bible. Often they refer to his power to cleanse us, however temporarily, from our sins. But, as we will discover in our discussion of faith and prayer, healing can also be physical in nature. God can heal us of our sicknesses, no matter how grave. He can do so because we believe he gave us life, he is in control of life, and he is beyond death. His power to heal our sickness is manifest as a miracle, since the act itself transcends any healing a human – not even the most gifted doctor with the most advanced technology – can achieve.

Where do we get this idea that God can work miracles? Other than the fact that our definition of God implies that as the most perfect human being God must have the power of healing, we can turn once again to more practical, and real experiences that provide us proof of God's miraculous healing powers.

As an adult, if you have been around toddlers who have begun to walk and run, you know that they bump into things, fall down, and hurt themselves. Rarely is the harm of any serious or consequential nature, but the infant is hurt and they come to you in pain, crying. What do you do? In most cultures around the world, we "kiss it and make it better," or something similar. It's a very simple act; a magic trick in a way. We marvel at how credulous our children are, to fall for such a trick, because they do – over and over in their infant years. They get up, go about their play, and immediately forget about the pain.

Now let us imagine ourselves as the child. The parent, or caregiver, or adult in the room, has just given us a miracle cure, instantaneous in its relief. How could such a person with the power of healing not be a God-like creature in our eyes?

Remember too, that these experiences are repeated time and again. They are lodged in our memory banks, inaccessible as surface memories, but powerful impulses that provide us with a foundation of belief that some entity called God must have the power of healing – the power to work miracles.

The Essence of God's Goodness

Let us now summarize what we know of the quality of goodness that exists in God. In American culture, it is common to say "God is good all the time, all the time God is good." This saying does not refer to the mundane qualities of goodness. God is not just good in the sense that today's weather is good as opposed to yesterday's gloom and cold. God must be good in a transcendental sense, beyond human capacities to be good. God's goodness must be of a sort that only our imaginary, perfect human being could have.

This exalted form of goodness reaches into God's innermost being, and only words such as *all-holy* and a*ll-sacred* can apply. Only an all-holy entity can never sin. From God's holiness comes his perfect love, and his capacities of mercy and compassion. Only an all-sacred God can perform miracles for us. And as God is heaven, and heaven is God, heaven must be the "most holy" place.

Now that we know what we mean when we say "God is good", and now that we have a definition for the good, or goodness, in terms of our objective here of understanding God, we must come to terms with the reality that *we ourselves will never have these qualities of Goodness.* We can aspire to be good, as best as is possible for us, and those who are especially graced with goodness as evidenced in their words and their actions, we may describe as saintly. Saintly people are those who best exhibit the quality of altruism – of making sacrifices for others at a cost to themselves. But these are only *approaches* to goodness. We can never arrive fully at the destination, because we are not God.

It is precisely this weakness, our sinful nature, which is brought to light when we contemplate God and wonder about his goodness. How do we deal with this sinful nature? This is where prayer and faith come in, the next topics on our journey of discovering the source of God-belief, and the role we wish it to play in our life.

CHAPTER SIXTEEN

What is Prayer?

Lord, empty me of me, so I can be filled with You.

- Traditional Christian prayer

A Return to Infancy

Prayer is the act of recreating the emotional state of our infancy, so that we may commune with the figure we identify as God. In its purest sense, prayer allows us to return to the security, love, and carefree existence of our infant and early childhood years, when we knew a literal heaven in the presence of God, who was represented by God-like figures. These figures were real people – our parents or our caregivers. As we grew older, we were given a replacement for these God-like figures, an imaginary individual called "God," who we could then craft into our own image of what an idealized human being would be like.

The God we pray to is Illusory God, but Illusory God is a component deity, in whom the most important figure for prayer purposes is Behavioral God – the third stage of development of our God-belief. Behavioral God is the God conditioned by our infant and early childhood introduction to language. Once we come to understand the language of our parents or caregivers – the God-like figures in our young lives – we understand more about the meaning and emotion behind their behavior. We understand more about why some behaviors are filled with negative intentions that can be frightening and judgmental.

The intrinsic relationship we have with this God is that of subordination. We are in every respect a lower creature, unable even to fully comprehend God's perfection, which is completely understandable, since such a thing as an idealized human being doesn't exist in reality. In the tradition of the Abrahamic religions, we are not just a lesser human than God, we are a base, foul, intrinsically sinful individual who is unworthy of God's love and must therefore strive perpetually to earn it. The love we seek is always just beyond our reach – there is always something God finds lacking in our character. In truth, God's full love for us will always be beyond our reach, because an idealization cannot love us, and we ourselves can never be perfect. The true love we seek is self-love, a healing of our psyche, sundered in two when we chose to believe in God, parceling him off as such a perfect being that we imagined he was no longer part of our mental makeup. We not just idealized him, we anthropomorphized and eternalized him, imagining God as a being existent in reality, physically located apart from us, but always available mentally for a conversation.

In prayer, we work to heal our human psyche, but we can only do so temporarily, and only during prayer. This is why prayer is so alluring; it gives us the illusion that psychologically we are united again with ourselves. Our self-loathing disappears, however briefly. The up-lifting of spirit that we often obtain from prayer is not an infusion of the Holy Spirit; it is nothing more than the power of positive thinking. This is why prayer is the ultimate form of self-deception. We truly believe we are talking to God, and he is answering us. As we have noted previously, neuroscientists have shown that when subjects are in prayer, their Temporal Lobe lights up in the exact same fashion as if they were talking to a third party. The difference is that in prayer we are doing nothing more than having a conversation with ourselves, which is by no means a bad thing and can certainly be a useful exercise without the interposition of an imaginary God. But the presence of God in this conversation ensures one critical thing: we must approach the conversation as if we were an infant or child. Participation in prayer is necessarily an infantilizing experience, because it was as an infant that we were truly at one with God, and it is as an infant that we can once again reunite with him, if only in a psychological sense.

Almost everything about the physical act of prayer is intended to reinforce our infantilized condition. We kneel, we grovel, we lie prostrate on the floor, and we hold up our hands in submission. One of the earliest forms of prayer in Christianity is the orant position, in which the hands are lifted above our heads, with the palms facing outward. The orant position is a classic depiction of the Christian asking as a child to be lifted up by the Father, and not a few hymns in Christianity plead that God "lift us up." Islam takes many of the different stances of prayer found around the world and incorporates them into the required daily prayer of *salah*. There are four postures taken and then repeated under *salah*: standing with head bowed and arms crossed; bowing from the waist with hands placed on the knees; kneeling at rest, meaning sitting on the back of one's feet; and finally prostration, where from a kneeling at rest position the worshiper bows deeply and touches their head to the floor. This posture is repeated multiple times as different verses are repeated from the Qur'an.

The language of prayer is also meant to put us in an infantilized state. Christians refer to God as the Father in heaven, which means that we humans are but as children in his presence. To achieve Oneness with God, and to return to the Garden of Eden, we must prepare ourselves by removing all the sin we have accumulated since being banished from heaven. We must be pure, innocent, and ready for the infinite love we knew in our infancy and early childhood. That is why the prayer cited at the top of this chapter – "Lord, empty me of me so I can be filled with You" – is very popular, and represents what theologians often call the most worthy form of prayer, which is prayer not for our selfish worldly needs, but for our moral improvement.

But note the essential aspect here that affects our moral and ethical character: in the perception of the believer in God, morals must come from without and not from within. Goodness (or grace) is a gift of God, and it is one of his most precious gifts, because he is all-holy. By granting us the gift of goodness, God is reassuring us that we are worthy of being loved. This is another aspect of belief in God that we mentioned earlier: to believe in God is to allow our moral self to atrophy. Since in the believer's eyes goodness can only ultimately come from God, the believer waits for God, or his Regents, to explain what is good, and how it differs from evil. This is why believers in God often panic when in the presence of an atheist; they cannot fathom someone living their life without a Regent telling them what is good, and how to live their life in a purposeful way.

Another favorite saying of Christians is that "we have a hole in our heart that only God can fill." A popular American Evangelical preacher expressed this conviction well in one of his internet sermons:

> **The Bible tells us that there is a God-shaped vacuum in our hearts. There is a hole in our hearts that only God can fill. We were made to be connected to our Creator. We were made to know Him and to be plugged into His power.**
>
> **Your heart is designed to contain God. God wants to live inside of you but when your life is filled with other things, there's no room in your heart for God. This means you're not plugged into God's power and that's why you're tired all the time. And that's why you're stressed out all the time and that's why you worry all the time.**
>
> **-From *Daily Hope with Pastor Rick Warren***

Pastor Warren is correct: His God does indeed want to live inside of us. In Pastor Warren's view, we long for God's presence to fill the hole in our heart, but the irony is that it is an absence entirely of our own making through belief in God. The struggle to fill this hole is life-long, because the quest is impossible to achieve as long as we accept the illusion that the purest aspects of goodness within ourselves are banished forever and lodged in the being we call God. No wonder we are "tired all the time," stressed out and worrying. It is exhausting participating in a perpetual psychological exercise of healing, unaware that we ourselves are the source of this stress.

Why do people do this? Why do believers subject themselves to self-loathing and an endless, futile exercise in self-healing through prayer? Think of it this way. Even if we purged the world overnight of all belief in God, and even if all religions were to disappear, there would still exist that part of our unconscious mind that maintains the conviction that we were once in the presence of a real God, and lived with him in the splendor and sublime comfort of heaven. *This was once our reality, and it is now our perpetual longing.* There are ways to satisfy this longing, and secular people with no belief in God find many useful means of living a "purposeful life" by confronting their unconscious longing for the purity, innocence, and stress-free existence of their infant years. Meditation, for example, is a secular equivalent of prayer, and provides as great a benefit to the non-believer as prayer does to the believer, which is why some religious denominations scorn meditation as a form of witchcraft.

We cannot, in the end, condemn those who believe in God by describing them as foolish or delusional. Their solution to our perpetual longing for a God-like presence in our life is to create such a God through mental gymnastics, and those who do this obviously derive a significant comfort in that act, or billions of people over human history would not have sought out God, or the gods. For such people, if the price to pay for their belief in God is the psychological burden of self-loathing, they are willing to accept that. The burden is experienced only now and then, and can be assuaged through prayer, which is a mental exercise that provides a temporary reconnection with one's "better self," that is, with one's God. This is why prayer is such an integral and essential component of belief in God. This is why many Christians say an underrated aspect of the Christian Triune God is the Holy Spirit, that part of God who speaks to Christians during prayer.

The Holy Spirit is the intercessory aspect of God's being, and he who comes to those in prayer to fill the "God-shaped vacuum in [their] heart." He is God's messenger to humans, and indeed, in early Christian and Jewish thinking, the Holy Spirit was considered the creative aspect of God – God the Creator, who merely spoke, and through his word the universe came to be. As was said in the opening verses of the Gospel of St. John, "In the beginning was the Word, and the Word was with God, and the Word was God." God spoke, and the world came into being, along with all things of this world, including man.

In what language does the Holy Spirit speak to us as believers? Obviously in our own language, since prayer is a conversation we have with ourselves. But in this conversation, are we as a believer in God summoning up our true, inner Goodness – recovering that which we assumed was lost to us when we were taught or came to believe that God actually existed? Yes, this is part of what is happening during prayer, the part of prayer where we do connect to our "better self," because the good elements of our being we thought we had cast away as the God of our creation, have never really left us. This is the part of prayer where we turn to our natural inclinations toward altruism; to doing that which we prefer others would do to ourselves.

But as with anything to do with God, the Holy Spirit may speak to us in our language, but his words are processed through many different cultural filters and beliefs. If we were Europeans or white Americans living in the late 18th century, we would more than likely commune with a Holy Spirit who would never once mention that slavery of the black man was sinful. It wouldn't occur to the Holy Spirit of that day and age to suggest such a thing to a white person engaged in praying to God, because the white culture of the time did not believe such a thing. To an African-American slave at that time, who had been given an education in the Bible by her master, the Holy Spirit would say something to her quite different, more than likely a reminder to read once again about the ancient Hebrews when they were in bondage to the Egyptian Pharaoh, and how someone named Moses rose to free them from slavery.

The Holy Spirit always speaks to us in terms consistent with our cultural norms. At this very moment in American history, the Holy Spirit is speaking words of doubt to many Christians regarding the sinfulness of homosexuality. The cultural norms in American society are changing, and the Holy Spirit is struggling to adapt, and in so doing, the Holy Spirit is starting to suggest to many who are praying over this issue, that a strict interpretation of the condemnation of homosexuality as found in Leviticus chapters 18 and 22, may not be appropriate or at all consistent with Christian charity and Christ-like behavior.

The most important filter through which the Holy Spirit speaks is that of one's religious upbringing. After someone of earnest intention has emptied themselves of themselves, by suppressing their ego and all assumed arrogance that goes with self-esteem, and when someone has then begun to pray to God, the Holy Spirit will almost always answer in terms that were set down by one's Regents. Many times the answer is a beautiful response, consistent with the messages one has received through a lifetime of religious education. The Holy Spirit may say, "It is wrong to spread gossip about someone at work," or "Christ-like behavior requires that you devote some of your time and resources to helping the poor." This is the moment when the Regent's belief in goodness, as reflected in hundreds of sermons about the importance of reaching out altruistically to others, bears real fruit.

This is why we cannot denigrate someone who practices prayer, just as we would not denigrate a nonbeliever who practices meditation or uses other techniques for what is called "centering" oneself. This term, in fact, is one of the most important forms of religious prayer – a prayer for centering oneself and coming to a closer understanding of one's inner thoughts and motivations. In the next section, we will cover briefly several of the most common and useful forms of religious prayer. This is a discussion of importance to all of us, whether we believe in God or not. We all have an inner existence that is the core of our being, and we spend most of our time with an internal discussion going on in our minds as background to our daily existence. That inner conversation that we constantly experience, which is an essential aspect of our human consciousness and self-awareness, is a form of prayer. The fact that religion has co-opted and normalized these inner conversations as integral to God belief should not surprise us in the least, since it is common for all of us to "pray," to have a conversation with ourselves.

The Religious Forms of Prayer

The major religions of the world do not agree on the forms of prayer to use in their worship. In fact, it is not possible for them to agree on such a thing, since religion is always filtered through local cultural norms. What a Buddhist in India desires out of prayer is a very different thing than a Buddhist in Japan would desire, and what a Japanese Buddhist raised in the Shingon sect expects from prayer is different still from a neighboring Buddhist who is a member of the Rinzai-zen sect. But all believers, whatever their religious background and beliefs, approach God from essentially the same place – the unconscious memory of him from their infant years. Consequently there will always be some similarity to religious approaches to prayer, as prayer tends to parallel our relationship with God as an infant. Let us look at seven such similar approaches to prayer.

The Centering Prayer

Centering prayer is the least "prayerful" of the different disciplines of prayer, in that specific words are not said to oneself. Quite the contrary, centering prayer attempts to empty the mind of deliberate thoughts, and allow the practitioner, sitting in a place of quiet and alone, to develop a stream of consciousness that brings them closer to the God within them. As such, it is closest to meditation, especially as practiced by Buddhist monks and nuns. In Christian terms, it is a prayer, according to one Catholic catechism, "in which we experience God's presence within us, closer than breathing, closer than thinking, closer than consciousness itself. This method of prayer is both a relationship with God and a discipline to foster that relationship."

The centering prayer is one of the oldest forms of Christian prayer, practiced as early as the 2nd century by monks living alone in the Egyptian desert. While Christians do not recite a formal prayer when in contemplation, not even the Lord's Prayer, they can select a single word to repeat to themselves if they find their mind wandering on mundane, material matters. This is known as *lectio divina* (Divine Word prayer), and is a practice used regularly by Benedictine monks. It is also closest to the Buddhist practice of chanting "Om" during prayer.

The stated goal of those using centering prayer is to feel God's presence within themselves, without thinking specifically about that presence or attempting to bring it about consciously. This is why centering prayer is an example of the desire to "empty myself of myself," allowing God's presence to enter oneself. It is a passive practice that is designed to remind oneself of the sense of the sublime, and of the perfect love one experienced as a helpless infant in the presence of supremely powerful God-like parents or caregivers.

Within certain traditions of all the major religions, the ultimate goal of centering prayers is to obtain Oneness with God. Those who reach such a state, and again they cannot force this upon themselves, as it must come from God, say that it is possible to lose all sense of physical boundaries, and achieve an understanding that everything in the universe is connected by a immensely powerful sensation of love. This force of love – of perfection and endless bounty – is the "ground of all being" that theologians discuss. It is God as the Creator, standing apart from his creation, which was born out of an act of divine love. With those people who reach this state of Oneness with God, there is no need to see God visually, though many mystics over the ages have reported visions of God, Jesus Christ, the saints, heaven, and even hell. The God of Oneness is said by believers to be invisible but his presence is so overwhelming that it is not necessary to understand where he is or what he looks like – he is everywhere, within everything and everyone.

The experience of Oneness is very real, and not something that we must accept simply by the words of the prayerful. Oneness can be scientifically established, as we saw in earlier chapters, and is referred to as ego dissolution by neuroscientists and psychologists.

The Prayer of Supplication

Supplication is the act of asking for something one desires. It is the most common form of prayer by many accounts, and this may be related to the fact that it is the one thing a helpless infant does most frequently – cry for food, or attention. The Lord's Prayer is a prayer of supplication: "give us this day our daily bread." This statement is a reflection of our infant condition, a belief that all of our earthly wants come ultimately from God. Jesus urged his followers to cease their endless fretting over where their material wants would come from. "And why do you worry about clothes? See how the flowers of the field grow. They do not labor or spin."(Matthew 6:28). St. Paul reminded his followers of the same thing: "Be careful for nothing [don't worry about anything]; but in everything by prayer and supplication with thanksgiving let your requests be made known unto God." (Philippians, 4:6). Both Jesus and St. Paul were masterfully tapping into our unconscious desires to be children once again, embosomed in the safety provided by our parents or caregivers.

These exhortations by the two founders of Christianity are very important. They describe the true benefit of the prayer of supplication. The benefit is often thought to be that if you pray for something specific, especially something material, like a new job or a winning lottery ticket, the Lord will provide. Indeed, many people approach prayer with exactly that intention – "give me something, O God, which I need!" It is easy to ridicule a religious person for turning God into Santa Claus, and many Christians urge their fellow worshipers to avoid such selfish prayers. This ignores the fact that many who pray in supplication also add the provision, "thy will be done," which is taken directly from the Lord's Prayer. In this way of thinking, God will provide us what we need, but not necessarily all that we want.

With this provision, the prayer of supplication returns its focus on to its real benefit: the ability to transfer the believer's earthly, material worries and fears on to God, in the manner of psychological transference referred to in Chapter Seven. This is what Jesus Christ and St. Paul were both suggesting: that prayer is a means of shedding one's anxieties over material things, and passing this burden along to God. This is the great benefit of such a prayer – the supplicant can walk away from their anxieties, at least temporarily, and rest comfortably in the knowledge that a supernatural force in their life will provide for them. They can be like infants again, for a short while.

None of this is to say that a Christian is absolved from the need to go searching for a job or otherwise take care of themselves. It is possible, however, within certain religious traditions, to reach a state of devotion that allows the individual not to have a job at all. The Buddha urged his monks to attain such a state by foregoing life's pleasures that derive from money and the desire for material possessions. The less one owns - the fewer ties one has to the material world – the better one is able to face death with little fear. In the Theravada discipline of Buddhism, monks practice the daily ritual of "pindapata," which is the act of approaching others in the community with a request for food by holding out a small, empty bowl. There are even some Buddhist monks who are forbidden by their order to feed themselves; all food must be obtained as a gift from others, and because of this many monks have a rule that they shall not eat after noon, so that they become used to eating sparingly in expectation that on some days there will be no food for them at all. The Buddhist bowl has become an important symbol of the religion, and pindapata is a practice that is a test of the monk's faith that providence will provide sustenance, as it did for the Lord Buddha, who began this practice. It is also a reminder that a life of complete dependence on others for sustenance did exist in our distant past, and can exist once again on earth and in the afterlife.

One of the greatest criticisms of Christianity, and religions in general, is that prayer doesn't work any better than expecting something to happen by chance. These criticisms focus almost exclusively on prayers of supplication, as if this is the only type of prayer participated in by religious believers. It may be the most popular choice of prayer, and certainly many Christians ask for things that are personal and could well be described as selfish. Such people are easy targets of criticism, but unsurprisingly, they will continue with their personal requests no matter how much they are criticized. Nor do they pay attention to the scientific studies – and there are a number of them – which establish that prayer does not work any better than chance. Someone who has spent a lifetime in prayer knows the success rate of prayer is 50% or less, and therefore never expects God to provide most of the time. Religious people have a readymade answer for such criticisms: "God gives us what we need, but not necessarily what we want," and "the Lord works in his own time and one must be patient when asking for help from God." This allows the believer to ignore all the many times God has not answered their prayers, and concentrate on those times God has seemed to answer their prayers. God gets all the praise for being good, and none of the blame for failing to respond.

Besides, the use of prayers of supplication is not similar to the way one would use a magic lamp. Prayer's benefits are mental and emotional, and with enough training and practice, a believer can come to understand that for them, prayer works almost all the time, because of its mental and emotional benefits. That is why people keep returning to prayer, and why it is such an essential element of God-belief.

The Prayer of Intercession

The prayer of supplication need not be a request for something for oneself. In its nobler form, it can be a request for something that would benefit others. Then it becomes a prayer of intercession, a request for God to intercede in someone else's life and satisfy one of their needs. Here is a typical example from the website of Joel Osteen, a prominent American televangelist:

> **Calling ALL prayer warriors. We made my husband Josh a doctor's appointment for this Friday and I'm asking for your prayers that the doctor clears him and doesn't refer him to any doctors and surgeons and that he doesn't want him to have any more testing done and that he says he is fine and that he doesn't want him checked out by any more doctors or surgeons either. Please pray the doctor says that he doesn't have a hernia and that the lump on his chest is normal and nothing serious or life threatening and that the doctor says he can return to normal activities. In Jesus's name we pray, Amen! Please keep him in your prayers daily, I am so worried and stressed.**
>
> **- Facebook.com/JoelOsteen**

Notice that the supplicant here, who is calling for a prayer on behalf of her husband, is emphasizing the stress that accompanies this situation affecting someone she loves. The act of praying for intercession from God is an act that can reduce this stress, and notice further that the petitioner believes that the more people who pray for help in her time of stress, the more likely it is that God will answer favorably. She starts her prayer with "Calling ALL prayer warriors." This is both a common and traditional approach to prayer in Christian communities. If someone is ill, the entire church will pray for that person to be healed. The internet has expanded the community of the prayerful enormously. In the case of this petitioner, she is appealing to a minister who has over 10 million followers on the internet.

Prayers of intercession can be far less personal, and deal with grave issues of justice and peace. This example is from a Catholic hymnal:

> **Father, you have given all peoples one common origin. It is your will that they be gathered together as one family in yourself.**
>
> **Fill the hearts of mankind with the fire of your love and with the desire to ensure justice for all. By sharing the good things you give us, may we secure an equality for all our brothers and sisters throughout the world. May there be an end to division, strife and war. May there be a dawning of a truly human society built on love and peace.**
>
> **We ask this in the name of Jesus, our Lord.**

This prayer has two levels of intercession operating. The first is on behalf of the petitioners who seek "justice for all" and an end to strife and war. The second level of intercession is found in the last line: "We ask this in the name of Jesus, our Lord." Christianity has a complex definition of God through its doctrine of the Trinity. God the Father is in theological terms equivalent to God the Son and to the Holy Spirit. All three are aspects of the divine nature of God. But they are also independent personalities, at least in terms of the Father and the Son, which as a doctrine speaks directly to our creation of God from our experiences as children in the presence of true god-like figures. God the Father is the Creator, the fearsome God (Yahweh) of the Old Testament. God the Son, that is Jesus Christ, is the ameliorating God. His role is to stand between the believer and God the Father, interceding on their behalf with the God who can grant requests, because God the Father is God the Creator. Jesus Christ can also grant requests, especially if they deal with health issues, since Jesus was a healer during his lifetime. As with prayers of supplication, prayers of intercession work half the time, at best, but when they do, the believer finds ample reason to offer a new type of prayer: that of thanksgiving.

The Prayer of Thanksgiving

In its simplest form, the prayer of thanksgiving is offered unto God for a particular favor he has granted the believer. If the woman in the example above discovers that her husband no longer needs to consult other specialists or surgeons, or better yet, is cured of the lump in his chest, she has a bountiful reason to thank God. In doing so, she is acknowledging something that she learned in her infancy: she is helpless in the face of life's basic misfortunes, and at such times, only a God-like figure can come to her assistance. A prayer of thanksgiving reestablishes her subservience and dependence on God.

This prayer is also an example of the power of positive thinking. When a believer gives thanks not for something specific, but for all that they have, and especially for personal benefits such as their health, family, and even their very existence, they are lifting up their spirits. When they give thanks for the beauty of the world, or when they remind themselves that others have worse problems than they do, and that God will never burden them with something he knows they cannot handle, they are restoring a sense of equanimity and calm to their being. Here is a writer who captures perfectly the emotional benefit of a prayer of thanksgiving:

> **Being grateful dispels fear and anxiety. If we allow our minds to think about and celebrate all the good things we have (and things that we may well take for granted), then there is less room for negative thoughts in our lives. For example, the next time you are in the supermarket, why not thank God for the great abundance you are surrounded by? "Father, thank you for all the many wonderful foods that are here!" As you are at the checkout, take a moment to thank God in your heart for the items that you are packing – 'God, I really am so grateful for these things.'**

Believers can, and as a believer they are often encouraged, to go beyond merely thanking God, however beneficial that may be for their mental health. God deserves more. God-belief and religion demand that the believer renders unto God worship and adoration.

The Prayer of Adoration

This form of prayer is often ridiculed by non-believers. What sort of God requires that humans worship him constantly? If he truly were God, he would have sufficient self-esteem to love his creation – mankind – without demanding incessant sacrifices and tributes. The extent to which those who believe in God seem to debase themselves as part of their belief is, indeed, a strange manifestation of God's love when looked at from outside of the religion, and from the point of view of the non-believer. Our purpose here is not to say how ridiculous this is, but to question *why* this is so important to the religious believer, because clearly some form of self-abasement is occurring when one believes in God.

Here is one pastor from the American Evangelical tradition, describing the importance of worship and adoration of God. The Bible citations are included in this quote for those readers who want some scriptural evidence for the arguments being made here in favor of adoration of the deity:

> **Praise creates humility before our Heavenly Father (1 Peter 5:5-6). We see how great and mighty God is, we learn how dependent we are on God and we cross over to learn that we need to depend on Him for every aspect of our lives. We surrender our lives totally and we become dependent on God for every aspect of our lives.**
>
> **Adoration of God breaks the strongholds of the sin of self-righteousness. Praise takes us to a place of brokenness and true repentance as we see ourselves for who we are in God's Eyes (1 John 1:9).**
>
> **When we adore God we focus on the holiness of God and how perfect God is. We see our sin in a new light and at a deeper level. Praise opens up our hearts and impacts our willingness to listen to God (1 John 5:14).**
>
> **Many times we have an agenda when we approach God. We ask God according to our will and agenda not His. Adoration fixes our faith on God's Wisdom, His Goodness, and His Power.**
>
> **As adoration of God increases, the more submissive we are to His Will.**
>
> **-Riggedreality.net**

Notice the heavy emphasis in this discussion on words like "humility," "depend," "surrender," and "submissive." In focusing attention on the perfection and holiness of God, "We see our sin in a new light and at a deeper level." Perhaps the most significant line is the one which says that adoration of God takes the believer to a place of "brokenness and true repentance," and that believers will see themselves "for who we are in God's Eyes."

How then, does God see his worshipers? Apparently not as individuals worthy of unconditional love. God's love is highly conditional. If we follow through on the scriptural quote of 1 John 1:9, we find:

> **If we claim to be without sin, we deceive ourselves and the truth is not in us. If we confess our sins, he [God] is faithful and just and will forgive us our sins and purify us from all unrighteousness. If we claim we have not sinned, we make him out to be a liar and his word is not in us.**

God's love is conditional on the acknowledgement of the believer's deeply sinful nature, which is due to the inheritance of Original Sin, but due as well to the simple fact that if someone believes in God, they have imbued him with complete holiness, and left themselves psychologically with an imperfect mental makeup. When the pastor in the quote above says that adoration reveals "we need to depend on Him for every aspect of our lives," he is bringing his readers back to their infant state, when they truly were dependent upon God-like figures. The prayer of adoration is the most infantilizing form of prayer, with its special emphasis on the weakness, complete dependence on God, and inherently unholy nature of the believer.

So far, the process of describing the different forms of prayer has taken us from centering, to supplication, to intercession, and then to thanksgiving, all of which follow an upward path of emotional satisfaction. The believer's satisfaction increases as they explore each of these forms of prayer, until they reach the pinnacle – the prayer of adoration. Here, they are now within the sublime presence of God. They are back in heaven. But there is a darkness there, a reminder of their complete helplessness as infants, because to adore God fully, they must return to an infantile emotional and mental state. In this state, they are also reminded of their unworthiness, and their unholiness, when compared to God.

From the pinnacle of prayer, which is the prayer of adoration, the believer now begins a downward journey in the emotional impact of the different forms of prayer. First, they must confess to their sinful nature. As was said in 1 John, if they do not confess their inherent propensity to sin, they are making a liar out of God.

The Prayer of Confession

The prayer of confession, also called the prayer of repentance or contrition, is not simply a statement of a believer's sins; it is a plea for forgiveness. Acknowledging one's sins is one thing; sincerely desiring to sin no more is much harder, yet essential if one is to be restored to God's grace, and receive his mercy. Here is an example of a complete acknowledgement of one's lowliness and unworthiness, coupled with a true desire to be forgiven:

> **O Allah! You are my Lord; I am a slave.**
> **You are the Creator; I am the one created.**
> **You are the Provider; I am the one provided.**
> **You are the Owner; I am the one owned.**
> **You are the Mighty and Glorious; I am the one abased and wretched.**
> **You are the Absolutely Rich One; I am the one absolutely poor.**
> **You are the All-Living; I am the one dead.**
> **You are the All-Permanent; I am the one mortal.**
> **You are the All-Munificent; I am the one miserly.**
> **You are the All-Benevolent; I am the one doing ill.**
> **You are the All-Forgiving; I am the one sinful.**
> **You are the Grand One; I am the one despicable.**

You are the All-Strong; I am the one weak.
You are the Giver; I am the one begging.
You are the One Giving Security; I am the one in fear.
You are the All-Generous; I am the one in utmost need.
You are the One Answering pleas; I am the one pleading.
You are the All-Healing One; I am the one sick.
So forgive me my sins and spare me and heal my ills, O Allah! O All-
Sufficing One! O Lord! O Faithful One! O Most Compassionate One! O
Healer! O Munificent One! O Restorer to Health! Pardon all my sins,
and restore me to health from all illnesses, and be pleased with me for
all eternity! Through Your Mercy, O Most Merciful of the Merciful!

Islam has many interesting prayers, but this is one of the most eloquent, with its citation in capital letters of some of the 99 descriptions of Allah, and then the juxtaposition in small letters of the opposite, feeble condition of humans. Having admitted to so many weaknesses and sins, the writer ends with an extended plea for understanding and compassion. Of the three Abrahamic religions, Islam more than the others puts great stress on God's mercy. The writer throws himself upon the mercy of God, in the last line, describing Allah as "the Most Merciful of the Merciful."

Religion, having obliged the believer to explore the darkness of the soul, requires next that they confess their sins to God and beg his forgiveness. This is a form of supplication, a prayer not for something material in their lives, but a prayer for something spiritual – an acknowledgement from God that he does indeed forgive all mankind and loves his worshipers, unworthy and guilt-ridden creatures though they are.

This leaves one last, unexplored form of prayer – the prayer of imprecation.

The Prayer of Imprecation

This prayer, sometimes called the prayer of interdiction, is an intercessory prayer, except the believer is not asking for something beneficial for someone else. She is asking for something harmful to another. The prayer of imprecation is a form of religiously sanctioned curse. Nowhere within Christianity are prayers of imprecation more frequently invoked than in the Book of Psalms, presumed to be written by King David. Here is but one example:

> **Break the teeth in their mouths, O God Lord, tear out the fangs of those lions! Let**
> **them vanish like water that flows away; when they draw the bow, let their arrows**
> **fall short. May they be like a slug that melts away as it moves along, like a stillborn**
> **child that never sees the sun.**
>
> **-Psalm 58**

The Psalms, many of which are highly poetic and gentle, are filled with dozens of these imprecatory verses that make for difficult reading. It must be remembered, however, that anything to do with belief in God and religion, including scripture, comes to us through cultural filters. The condition of the Hebrews at the time of King David was that of a nation under threat from several neighboring enemies. Yahweh had, among his many qualities in the Old Testament, the quality of martial victor and avenger, who worked on behalf of the Israelis to defeat their foes. We don't know the full historical facts regarding the Hebrews of the Old Testament, but we do know that many books of the Old Testament speak of warfare, conquest, and vengeance as a normal feature of Hebrew life at the time. We should not be surprised that the Psalms reflect this cultural concern with a Tribal God of vengeance.

Jesus Christ's great contribution to Judaism was the reinvention of Yahweh. There were already trends under development within the Jewish community in the 1st century BCE to move away from the puissant God of the Old Testament to someone more "civilized," more in keeping with the fact that hundreds of thousands of Jews were living peaceably in cities around the Mediterranean – from Antioch to Corinth to Alexandria and to Rome – and had long since abandoned the agricultural and nomadic life of the ancient Hebrews. Jesus introduced the Jews to a revolutionary new God – Abba, or God the Father. This was not a concept with which Jews were familiar. In the Old Testament, Yahweh has many styles or names, but God the Father is mentioned only around ten times in the entire Old Testament.

The idea of God as a father figure was essential to Jesus' understanding of God and of his relationship with God. To the Jews of his time, it must have sounded strange for someone to say that their prayer to God should begin, "Our Father, who art in heaven." Yet the Jews were ready for just such a personal, intimate, forgiving God, as were so many others in the Roman Empire. In due course, Judaism adopted more prayers devoted specifically to the Father, or Avinu.

Many others besides Jews were susceptible to viewing God in a new light, as a father figure. This is one reason why Christianity spread so quickly through the Empire. Christianity therefore had lesser need for prayers of imprecation, though they were invoked from time to time. In keeping with Christian dogma, however, prayers of imprecation were especially suited to damn one person in particular – Satan. In this sense, they served a proper theological purpose, since the battle against the Evil One was a paramount consideration in Christianity. Here is a part of the famous prayer of interdiction against those possessed of demons, as written by St. Basil the Great of the Orthodox Church:

> **O God of gods and Lord of lords, Creator of the fiery ranks, and Fashioner of the fleshless powers, the Artisan of heavenly things and those under the heavens, Whom no man has seen, nor is able to see, Whom all creation fears: Into the dark depths of Hell You hurled the commander who had become proud, and who, because of his disobedient service, was cast down from the height to earth, as well as the angels that fell away with him, all having become evil demons. Grant that this my exorcism being performed in Your awesome name, be terrible to the Master of evil and to all his minions who had fallen with him from the height of brightness. Drive him into banishment, commanding him to depart hence, so that no harm might be worked against Your sealed Image. And, as You have commanded, let those who are sealed**

receive the strength to tread upon serpents and scorpions, and upon all power of the Enemy. For manifested, hymned, and glorified with fear, by everything that has breath is Your most holy Name: of the Father, and of the Son, and of the Holy Spirit, now and ever and into ages of ages. Amen.

St. Basil the Great lived in the late 4th century CE, at a time when Christianity was solidifying into its theology the concept of cosmic temporal Dualism. A prayer of imprecation against Satan has one great advantage over other prayers: no one has any way of knowing if it has been successful or not. This is not so with most other prayers, especially those of supplication, which scientific studies have shown are never successful more than half of the time. What else can we learn from science about the efficacy of prayer?

What Modern Science Teaches Us about Prayer

Does prayer work? This question has been looked at repeatedly by groups that have varying degrees of interest in the outcome. One traditional way of analyzing the problem has been to ask a group of religious people to pray for a very specific outcome, such as an increase in personal income during the next two weeks, or a chance meeting with a long lost friend. No amount of prayer has been shown to produce an outcome that would be better than a 50% possibility.

But that is analyzing prayers of supplication. What about less-selfish prayers; intercessory prayers, for example? The religious and spiritual organizations which have sponsored studies on the efficacy of prayers of intercession have tended to find that prayer "works," but unfortunately the studies themselves are flawed and prone to providing a particular result. The organizations which conducted these studies were typically interested in either religious prayer, or spiritual remote healing, so the study would focus on individuals undergoing medical treatments for illnesses such as cancer. The problems with these studies was that they included too few subjects to provide meaningful statistical results; the results were too broadly defined and allowed for any number of interpretations (the patient was deemed to have benefited from prayer if they simply "felt better"); the outcome was determined by the patient rather than by an outsider; and it was impossible for a third party to duplicate the test and produce the same outcome.

When independent organizations conducted their own studies and had no interest one way or the other in proving prayer and remote healing worked, the results were definitive. Duke University conducted its own study of 748 patients with heart problems, and in 2005 published the results in *The Lancet*, Britain's leading medical journal. The patients were divided into four categories: those who were assigned people to pray for them from a distance (this was unknown to the patient); those who were given MIT therapy (music, imagery and touch); those who received both MIT therapy and distance prayer; and those who received neither of these "benefits." There was no statistical difference in medical outcomes among any of these four approaches, so clearly prayer, or remote healing, did not work.

At the same time, Harvard University conducted its own study, titled *Study of the Therapeutic Effects of Intercessory Prayer*. Harvard coordinated among five other hospitals to study 1,802 individuals, all of whom were to undergo standard coronary bypass surgery. The study was partially funded by the Templeton Foundation, which specializes in research on religion. Various church congregations provided third parties who were willing to pray anonymously for strangers about to undergo surgery. The patients were divided into three groups: those who received no prayers and were not informed of the study; those who were informed of the study and were told beforehand they may or may not be prayed for (even though all were prayed for anonymously); and those who were informed of the study, told they would receive prayers anonymously for two weeks beginning the day before the surgery, and did receive prayers as promised.

In reviewing the number of patients who had cardiac complications post-surgery, including severe complications such as stroke and death, the study team found there was no better outcome for those receiving prayers. The press statement from Harvard announcing the study results said:

> **In a clear setback for those who believe in the power of prayer, their prayers were not answered. Prayers offered by strangers did not reduce the medical complications of major heart surgery. Not only that, but patients who knew that others were praying for them fared worse than those who did not receive such spiritual support, or who did but were not aware of receiving it.**

Why would prayer actually worsen the outcome for these patients?

> **We found increased amounts of adrenalin, a sign of stress, in the blood of patients who knew strangers were praying for them. It's possible that we inadvertently raised the stress levels of these people.**

One theory put forward by the study team was that patients, when learning people would be praying for them, might conclude their medical condition was worse than they thought. In any event, because of this outcome and the ethical question involved of stressing test subjects on matters as serious as the outcome of major surgery, no hospital has attempted to duplicate this study. Religious groups still insist that they have known of cases where intercessory prayer has benefitted people, and spiritual groups still insist that remote healing is possible because our knowledge of how human consciousness is linked is still very primitive. That may be so, but the Harvard University study opened up a new line of inquiry, which is the real possibility that prayer can raise the stress levels of those receiving the prayer, and thus cause damage. Prior to this result, the debate had always been over whether prayer worked or not. Now the debate has begun to shift, into whether prayer can damage people under certain circumstances.

Faith as the Handmaiden of Prayer

It is frequently asserted by religion that prayer cannot work without faith in God. One must have a sincere belief and trust in God for him to even consider your prayer. This is a very convenient argument, since prayer works only as a matter of luck or happenstance. Every time prayer does not work, the believer has reason to blame themselves for this failure, and not God. Perhaps their faith isn't strong enough. Perhaps if they believed in God more, in his power to work miracles, and in the soundness of his plan for their life, then miracles would occur more frequently. So they are told.

This is one of the ways faith works. It exploits God's failures to give what one needs in life, by shifting the focus of these failures on to a believer's own weakness. "O ye of little faith!" said Jesus repeatedly to his disciples and other followers. We will explore in the next chapter the importance of faith.

CHAPTER SEVENTEEN

What is Faith?

To one who has faith, no explanation is necessary.
To one without faith, no explanation is possible.

-Thomas Aquinas

A Bulwark Against Doubt

Faith is the practice of establishing psychological defenses within one's mind, so that a person who believes in God has little or no reason to doubt that belief. Faith's most faithful ally is that part of an individual's unconscious mind, which knows that God and heaven once existed, and which desires them to exist once again. The adult, mature mind, however, is in possession of the truth. It knows our existence is anything but heavenly, and that no God exists in the reality that we inhabit, or grants us a reprieve from death in an "afterlife." The truth is that death comes to us all, and it leads to the same oblivion we inhabited before birth. The rational mind rebels against ideas which do not conform to this reality, such as the idea that some people have escaped death and returned to life on earth. Faith is designed to hide the reality of our finite existence, and like a form of psychological repression, it allows the believer the ability to avoid confronting death in all its cruelty and finality.

Faith has a second and very powerful ally, and that is a believer's parents, or caregivers, or community, which pass on to them when they are children, and at a most vulnerable age, the concept that God still exists and that they will cheat death after all, by joining him in heaven in the afterlife. This concept is a direct appeal to the unconscious mind and the urge to believe in God, and it is virtually impossible to resist as a child. Even among atheists, the number among them who rejected God-belief and religion when they were a child is few; most atheists reject God-belief as young adults, and not earlier.

The history of mankind is such that God-belief and religion have triumphed over the adult, rational mind. Theism, the consequence of which is the belief in God or gods, has for the recorded history of mankind been the default belief system of most people. It is only in the past few hundred years that significant numbers of people have chosen atheism, which is a learning process that leads to disbelief in God or gods. Even so, atheism is a minority practice today, though growing in popularity, and since it is a minority opinion, most people respond almost exclusively to the urgings of their unconscious mind. Most people have been trained to erect the various psychological bulwarks which constitute faith, and those defenses prove for the entirety of many people's lives to be impregnable. What are these defenses, and why are they so strong?

The Armor of Faith

Faith is often described as the belief in things which have no existence in reality. St. Augustine said, "Faith is to believe what you do not see; the reward of this faith is to see what you believe." But this is not the definition of faith; it is the *consequence* of faith. Faith itself is a man-made invention that consists of a set of mental exercises which ward off rational doubts regarding God's existence or the necessity to practice a religion. Faith is an invention of Regents, acting on behalf of their religion, and is always crafted and molded to fit the needs of a particular religion, or even of its Regent. In this regard, faith is *promulgated* by religions which are relatively comfortable with congregants who have doubts about God's existence and which are managed by Regents who are collegial in their pastoral practices. But faith is *demanded* by religions which promote scriptural mythology as theological fact, or which function as cults where the Regent requires absolute obedience. Faith operates, in other words, on a continuum, whereby more faith is demanded of believers if the religion is uncertain of the power of its dogma, or where a Regent has positioned himself as sole interpreter of its dogma. Thus, the fundamental responsibility of any Regent is not to teach religion, it is to ensure that their members of the congregation maintain a strong faith in the religion. Each congregant must be girded with the necessary number of mental defenses so that, on their own, they can ward off any doubts or challenges to their belief in God and their religion.

Let's look at some of the defenses which constitute faith in God and religion, remembering that the number of these defenses required of a believer, the intensity with which they must be accepted and applied, and which ones to use, depend on where the religion and the congregation involved fall on the continuum of faith.

Fear of Death

The fundamental emotional appeal of faith is as an antidote to a believer's fear of death. Fear of death is the product of the truth of our finitude, that existential anxiety from which springs all religion. Faith is the analgesic - the balm – that smothers these anxieties with the soothing reassurance that there is an afterlife. All of us have a built-in defense against the truth. We repress thoughts of death as quickly as we can, so we can go on with our lives without confronting the truth. But how much easier is it to repress thoughts of death when we have a ready-made weapon in the form of a belief in God and an afterlife. This weapon has been invoked from the very beginnings of Christianity, and continues to this day. Here are some examples:

Oh death where is they sting? Oh grave, where is thy victory?

-St. Paul, 1 Corinthians 15:55

No trembler in the world's storm-troubled sphere; I see Heaven's glories shine, And, *Faith* shines equal, arming me from *Fear*.

-Emily Bronte

There is nothing that wastes the body like worry, and one who has any faith in God should be ashamed to worry about anything whatsoever.

-Mahatma Gandhi

Resist your fear; fear will never lead you to a positive end. Go for your faith and what you believe.

-Rev. T.D. Jakes

Fear of God's Judgment

God, as a construction of our own mind, operates in an entirely human manner when he is angry. This is the Bad Behavioral God – the God with pronounced, negative human characteristics. He punishes those who offend him, and nothing offends him more than someone who does not believe in him. As was said in the Qur'an:

Those who reject Faith, and die rejecting - on them is Allah's curse, and the curse of angels, and of all mankind. They will abide therein [in Jahannam, or Hell]: Their penalty will not be lightened, nor will they receive respite.

-The Qur'an, Sura 2:161-162

To reject your faith is to reject your religion, which is to reject God. To reject God is to deny him power over you, and there is no greater form of rejection of God than to deny that he is a god, or that he even exists. This is blasphemy. Jesus was remarkably understanding about those who would disagree with him (as the Son of Man, or a prophet in Old Testament terms). But he did not tolerate those who denied God, whom he described as Abba, or Father, and sometimes as the Holy Spirit.

Therefore I say to you, any sin and blasphemy shall be forgiven men, but blasphemy against the Spirit shall not be forgiven. And whoever shall speak a word against the Son of Man, it shall be forgiven him; but whoever shall speak against the Holy Spirit, it shall not be forgiven him, either in this age, or in the age to come.

-Matthew 12: 31-32

This is why Christianity refers to blasphemy as the unforgiveable sin. The penalty for blasphemy is eternal damnation, which seems slightly logical since the blasphemer has already chosen to distance themselves from God. But as with anything to do with hell and eternal damnation, the punishment seems grotesquely disproportionate to the sin. How insecure must God be that he bristles at even the thought of someone denying him?

The answer to this conundrum is quite simple. God is not insecure in the least, because that which does not exist cannot be insecure. It is his Regents on earth who believe in him who are insecure, so they invent a hellish punishment for disbelief, and they threaten their followers and all non-believers with hell should they deny the Regent or deny their religion. As we have shown earlier in this book, this is a very powerful weapon, and as such it works in tandem with the appeal to faith. The threat of damnation, in other words, rests in the back pocket of the Regent should he need to use it, but it is far more pleasing for the Regent to use a carrot by offering the believer an addition to their arsenal of faith. They will say, "believe in God and you shall be among the elect; those who disbelieve shall be damned."

Fear of the Doubter

Some believers are able to live in a hermetically closed world in which they only meet right-thinking believers like themselves. Many others, however, come across non-believers, who appear in many forms: as convinced atheists, those uncertain as to whether there is a God, those who believe in God but do not trust religion or its Regents, and those who ridicule believers. Then there is the general culture, to be found in television, movies, pop entertainment, and the press, all of which represents a secular approach to religion that either ignores it or otherwise gently mocks it (unless it is profitable to produce something that panders to a religious audience). Finally, there is what can be called the "religious other," or that person whose approach to God is anywhere from marginally to radically different from that of the believer.

How does a believer maintain their own faith in the face of so much external opposition or expressions of doubt? They tell themselves that those who present religious views different from their own are lying, or they are ignorant, or they are benighted by false teaching. This defense springs up almost instantly in the face of opposing or questioning viewpoints; the believer learns how to quickly shut their ears to such talk. Adding one or two favorite scriptural passages often helps, and for added security, the believer can throw out a stronger defense, such as "You are going to hell!" This often works to silence the other party, if only by shock, and it helps the believer reinforce their faith, by reminding themselves of the consequences of doubt.

Where did this practice of tuning out opposing viewpoints arise? In Christianity, from no less an authority than Jesus Christ.

> **Jesus turned and said to Peter, 'Get behind me, Satan! You are a stumbling block to me; you do not have in mind the concerns of God, but merely human concerns.'**
> **-Matthew 16:23**

The Prophet Muhammad gave similar advice:

> **Therefore listen not to the Unbelievers, but strive against them with the utmost strenuousness...**
> **-The Qur'an, Sura 25:52**

Closing one's ears and minds to outside influences is a self-inflicted harm. But there is an even worse version of this defense: shunning the former believer who has lost their faith.

Fear of Estrangement from Family and Friends

Very few defenses of faith are more pernicious than the practice of shunning, sometimes called in some sects disfellowship. Shunning is reserved for apostates – those who re-nounce their previous beliefs. Shunning is not generally practiced in mainstream religions. It is found most frequently in smaller denominations and religious cults, especially where there is tight top-down control of the cult by a single, persuasive individual. This type of personality demands respect and obedience at all times to their requests and whatever doctrines they espouse (which can change regularly).

There are manifold aspects to the evil that is shunning. The person who is ousted some-times engages in behavior the outside world might consider to be insignificant, such as listening to popular music, or quietly questioning the theology of the leader. The price they pay is to be eternally separated from their parents, brothers, sisters, other relatives who are in the cult, and all the other members of the cult. Life in these cults is often very insular with little contact with the outside world. The apostate usually has few skills to succeed outside of the cult, and has no friends or family to rely on. They find themselves torn suddenly and permanently from the only family, and indeed the only life, they have ever known. The family within the cult feels the same sense of loss, but they are taught to suppress these feelings.

Another evil associated with shunning is that it reinforces the power of the leader, or board of elders, over the entire cult. It is a fear mechanism that promises severe personal conse-quences for anyone who openly expresses doubts about their faith. This allows shunning to create cohesion within the group, at the expense of any individual thought being allowed.

Shunning is practiced regularly by major cults, such as Scientology and the Jehovah's Witnesses, who are well-known for their disfellowship practices. Shunning is found fre-quently on a family basis, and sometimes on a church basis, in conservative congregations in the U.S., such as among Evangelicals, Fundamentalists, Charismatics, Pentecostals, Mormons, and occasionally among Catholics, now that American Catholicism has aligned itself formally with the political right. Religiously conservative families are prone to use shunning against their own children, including teenagers, if they are viewed as rebellious because they reject God-belief or religion. In addition, teenagers who come out openly as gay in such families often find themselves thrown out onto the street. In the U.S., gay teenagers who have been forced out of their homes by their own parents, account for over fifty percent of all homeless teenagers in the country. The incidence of suicide and suicide attempts among this group of teenagers is staggeringly high, and the lifetime psychological problems caused by shunning are only now beginning to be understood and chronicled.

There is nothing at all redeeming to say about the perpetration of evil that is shunning, all in defense of faith, and so we will turn to one other defense mechanism used to prop up faith.

Reliance on Holy Scripture

Religion either has a written text that provides a basis for the theological beliefs of the members, or a set of practices and sayings that are passed down from one Regent to another and which serve the same purpose as a written text. We will call this theological underpinning "scripture," which in the Abrahamic religions consist of the Torah, Mishnah and related writings in Judaism; the Qur'an, Hadith, and related writings in Islam; and the Holy Bible, works of the Doctors of the Church, and related writings in Christianity. We refer to all of this as Holy Scripture.

Holy Scripture is almost always susceptible to interpretation and outright wonderment at the stories, claims, and contradictions inherent in the scripture. The Bible has at least 10,000 translation errors that have arisen over 2,000 years of Christianity, which also complicates interpretation. Even if Holy Scripture were to remain perfectly coherent internally, passed down over thousands of years without a single change to the original document, scripture would still require interpretation, especially considering how many different human hands were involved in its creation.

For this reason, Holy Scripture is always a *received* and *learned* discipline for the believer. The believer is given the scripture by the elders, or their parents and other teachers, and the believer is required to learn the scripture, often at a very early age when memorization of a catechism is mandatory for receiving Confirmation, a bar mitzvah, or graduation from a madrassa. The business of interpreting Holy Scripture is in the hands solely of the Regents, and *all* religions have some form of Regent available for this most vital of religious responsibilities.

Why are religions so protective of Holy Scripture? For religion, Holy Scripture is that which is defended by faith. Holy Scripture is the filter applied by the religion to God-belief and a longing for heaven. It is not to be questioned by the faithful in any way. Therefore it is trotted out as a defense of the faith, which tends to appear contradictory to outsiders, who find believers constantly engaging in circular reasoning to defend their faith. The surefire argument a Christian will raise at the point of frustration with a doubter, is, "But the Bible tells me so!" The non-believer listens to such an argument with equal frustration. They will say to themselves, "How can this person possibly believe that I, an atheist, will be convinced by a Bible quotation, when I don't even believe in the Bible?"

But that is not the point. The Bible quotes that are frequently used by believers are not intended to win over an atheist in an argument about the existence of God. *They are intended solely to buttress the faith of the believer* in the face of doubtful statements and arguments from an outsider. Just as scriptural quotes are used to buttress the believer's faith, so faith is used to invigorate the believer's resistance to sin, an important element in the life of any religious believer.

The Burden of Sin

Sin is that personal behavior which is found to be objectionable by religious authorities. Sin is in the eye of the beholder, who is ultimately a Regent as protector of the Faith. Sin is, therefore, an entirely religious preoccupation. As was said by a noted American pastor and theologian:

> For a Christian to be a Christian, he must first be a sinner. Being a sinner is a prerequisite for being a church member. The Christian church is one of the few organizations in the world that requires a public acknowledgement of sin as a condition for membership.
>
> -R.C. Sproul

A belief in sin is the direct consequence of the psychological disunion discussed earlier – the requirement that the believer in God invest in God all goodness and holiness, leaving for the believer a psychology that focuses on their own personal unworthy nature. All manner of vice and wickedness can be ascribed to the unworthy nature of a believer in God, and many a believer has suffered considerably when contemplating their evil nature, no matter how small the sin.

> Let the sinner know that he will be tortured throughout all eternity, in those senses which he made use of to sin. I am writing this at the command of God, so that no soul may find an excuse by saying there is no hell, or that nobody has ever been there, and so no one can say what it is like...how terribly souls suffer there! Consequently, I pray even more fervently for the conversion of sinners. I incessantly plead God's mercy upon them. O My Jesus, I would rather be in agony until the end of the world, amidst the greatest sufferings, than offend you by the least sin.
>
> -From The Visions of Saint Maria Faustina Kowalska

> It does not matter how small the sins are provided that their cumulative effect is to edge the man away from the Light and out into the Nothing. Murder is no better than cards if cards can do the trick. Indeed the safest road to Hell is the gradual one - the gentle slope, soft underfoot, without sudden turnings, without milestones, without signposts.
>
> -C.S. Lewis

> Have you ever considered the deeper implications of the slightest sin, of the most minute peccadillo? What are we saying to our Creator when we disobey Him at the slightest point? We are saying no to the righteousness of God. We are saying, 'God, Your law is not good. My judgement is better than Yours. Your authority does not apply to me. I am above and beyond Your jurisdiction. I have the right to do what I want to do, not what You command me to do.'
>
> -R.C. Sproul

How surprised should we be that the "most minute peccadillo," the most innocent of lies, the briefest thought of lust, the slightest feeling of envy, is enough to weigh so heavily on the mind of the believer that they would prefer "agony until the end of the world" rather than offend the God of their imaginings?

The calculation a believer must undertake to weigh the heinousness of any sin is similar to the calculation Pascal made to define his wager. Heaven is an eternity of bliss; hell is an eternity of torment. When the stakes are infinity high or low, it is impossible to treat the wager in a normal fashion. Logic and calculated reason are thrown out the window. The believer invests the entirety of their being in seeking heaven and avoiding hell, and consequently *any risk* of hell is deemed unacceptable. The weight of even one atom of sin on the scale is enough to incur everlasting perdition, and so the slightest sin is deemed as grievous as murder or adultery.

Hence faith takes on a double duty. Faith not only supports a person's belief in God, it serves as protection against sin.

> **When all is said and done, the life of faith is nothing if not an unending struggle of the spirit with every available weapon against the flesh.**
> **-Dietrich Bonhoeffer**

God is said by his believers to be omnipresent. Those who do not believe in God will dispute this claim, but to the believer, God's omnipresence is also ominous. God sees all, and knows all. He can see into the deepest and most remote corner of your soul, and he can ferret out even your sinful thoughts. This is one reason why, *in the life of the believer, sin is omnipresent, because God is omnipresent.* God may be all-holy, but he is a perpetual reminder of all that is evil in you. Sin is therefore a burden of belief in God which never ceases to plague the believer. The non-believer rids themselves of this burden, and therein lies an inevitable source of frustration if not anger when the believer meets an atheist. The believer intuits that the atheist is much happier without the burden of sin, and the believer is jealous of this fact. Jealous, but also curious as to how someone could achieve a release from a lifetime of suffering.

What must it be like to leave entirely the world of God-belief and religion? To the believer, the idea evokes fear. Disbelief in God is blasphemy in religious terms; the greatest of all sins. The wall of faith erected at the frontier of belief in God and disbelief in God is the highest wall of all, the most difficult to scale, and the most impenetrable. Most believers go nowhere near that frontier, beyond which is presumed to be a precipice leading to perdition.

It is time for us, at least conceptually, to scale that wall and see what is on the other side. I invite even the most deeply convicted believers to join me so as to see the other end of the spectrum of life, where those who do not believe in God pursue their personal search for not-death.

CHAPTER EIGHTEEN

What is Atheism?

*It's a strange myth that atheists have nothing to live for.
It's the opposite. We have nothing to die for. We have
everything to live for.*

-Ricky Gervais

Dismantling Faith

Atheism is the act of resisting the inner propensity we all have to believe in God, and establishing in its place a mental attitude that is free of the burdens of self-loathing and fear of eternal damnation. Atheism is a personal process of discovery. The *product* of atheism is *disbelief in God to some degree or other,* and disbelief in his promise of an eternal afterlife with him in heaven. An additional benefit is the conviction that mankind does not require a belief in God or in religion in order to create a better world or achieve the benefits that religion can accomplish.

Theism, which is a process of faith that results in a belief in God and all that he represents, is the default position for mankind, and as such requires only a passive intellect and will. To attain a belief in God is almost always nothing more difficult than being born and inheriting the religion of one's parents or guardians. The cost of God-belief comes later, throughout one's life, as the believer also inherits guilt through sin, a fear of damnation, and the necessity of performing obeisance to Regents who impose manmade constraints on believers, in the name of religion.

Atheism is a denial of that which everyone else seems to accept regarding the existence of God, and is obtained through the active intellect and will. To achieve atheism is a mental struggle, the costs of which are incurred upfront, and the benefits derived later, through freedom from guilt, and through the opportunity to explore life from the humanistic and natural viewpoint, rather than from the supernatural perspective.

Atheism is not a single pathway leading ultimately to the goal of disbelief in the existence of God. Atheism is a *process* of disengaging from faith in the existence of God, and as such the word "atheism" must be understood to cover multiple paths to different destinations carrying some degree of disbelief in God. These would include doubt, agnosticism, and humanism as possible pathways of disbelief. The statement "I am an atheist" should therefore be read to mean "I am involved in a process of disengaging from religious faith and belief in God." By this standard, someone who doubts the existence of God is an atheist, an agnostic is an atheist, someone angry at being deceived by religion is an atheist, a secular humanist is an atheist, and so forth.

The Goal of Atheism

We have provided two consistent, abiding principles in this book which can inform us about the goal of atheism:

- God is our concept of the idealized human being.

- All of us, believers and non-believers alike, seek not-death as an answer to the existential problem of our mortality.

These two principles are often melded together, either accidentally by atheists, or deliberately by religious leaders who love to claim that atheists are nothing more than people who have replaced the worship of God with the worship of mankind. Or, to cite another misconception, "Atheism is nothing more than a religion for people who wish to hide their obsession with God." The author Terry Pratchett described this way of thinking in his book *Feet of Clay:*

> "No it's not!" said Constable Visit. "Atheism is a denial of a god."

> "Therefore It Is a religious position," said Dorfl. "Indeed, a true atheist thinks of the Gods constantly, albeit in terms of denial. Therefore, Atheism is a form of belief. If the Atheist truly did not believe, he or she would not bother to deny."

> — Terry Pratchett

In this interchange, Dorfl is employing a classic straw man argument – imputing to the atheist a belief which they do not actually hold. Now an atheist would agree that God is, indeed, our idea of the most perfect human being possible, because all people carry such a conception. We develop this idea from the moment we are born, when it comes to us as an emotional impression derived from our experience with our mother or other caregivers. This is the critical moment in our lives when, in the real world, it appears to us there exists someone of superhuman powers who created the world we live in, gives us life, and provides us with all of our needs. Some years later we discover the truth about our parents or caregivers, which is that they are anything but superhuman creatures. They are flawed individuals in a deteriorating body which is ultimately leading them to their death, just as our body is going to betray us as well. But we are very much attached to the powerful concept of an idealized human being ready at any time to protect us and provide for us, and we don't want to give it up, even though we understand our parents are not examples of the idealized human being. So we incorporate this powerful concept into an illusion – Illusory God – in whom we will put our faith for the rest of our life, if we so choose.

But to repeat: the idealized human being is an illusion. When we enter the process of atheism, governed as it is by the necessity to approach the world with logical thinking and reason, we must abandon illusions. We must discard the idealized human being as something we seek. It is not, therefore, the goal of atheists to aspire to the idea of a perfect human being. Atheists do not seek perfection in human nature and do not worship the idea of a perfect human being.

Where atheists and believers can come to some consensus is on the second principle, because it has legitimate aspirational qualities that are applicable to believers and non-believers alike. Atheism, like faith in the existence of God, is a process that helps us achieve not-death. But there is a fundamental difference in the two processes of faith and atheism. The goal of faith is to achieve not-death through immortality in the afterlife. The person of faith operates in a world of illusion, believes that the afterlife is real, and that it exists in our material world in the cosmos, or perhaps outside of the cosmos in the presence of God. In either case, the afterlife is interpreted as real enough to house our consciousness, if not our physical body, for all eternity.

Atheism allows for no such illusion. Atheism is grounded in logical thinking, reason, and rationality. There is no evidence for the existence of an afterlife. Near Death Experiences, where the dying see a bright light at the end of a tunnel, find Oneness with the universe or God, see their deceased relatives urging them on into the light, and so forth, are often cited by religious people as proof of the existence of an afterlife. What science shows is that Near Death Experiences are phenomena which occur in the human brain as a result of the shutting down of the body at the point of death. They are more than likely common to all people and seem to be a physiological way to ease the transition from life to oblivion.

There is no scientific, factual, reproducible evidence for an afterlife. When an atheist talks about seeking a pathway to not-death, they are discussing leaving tangible, positive contributions to mankind while they are here on earth. Their "not-death" reward will be the fact that these contributions will outlast them long after they have died. Moreover, while it would be nice for any atheist to achieve global fame and immortality for some historically significant contribution to mankind, the expectations of an atheist seeking not-death are much more modest.

It is enough for any atheist to be able to expect that they will have been personally appreciated by only the friends, family, and acquaintances they met during their life. It is enough for them to say that their acts of kindness, their charitable giving, their assistance to others who are struggling in their lives, their capacity to spread cheer and good will – in short, their altruistic contributions – changed other people's lives in some meaningful way. It is enough for any atheist to say that they will be remembered by two or three generations who follow them in life, after which even their name and what they did in life will be completely, irrevocably forgotten. An atheist accepts this because it is unimportant whether they are immortalized by those who follow. The true compensation for an atheist who lives a life of purposeful altruism is that such acts have unexpected, and sometimes profound benefits for future generations, extending over very long spans of time.

This is the secret to the willingness of atheists to embrace death, when it comes, as an inevitable part of Nature. So often it is said by critics of atheism that atheists have nothing to look forward to in life, and no purpose to their life. Everything to them is said to be meaningless because at the end of their life, there is nothing but darkness – oblivion. And atheist concurs that it is their *body,* and their *consciousness,* which enter into oblivion. But their altruistic contributions to others, however insignificant they appeared at the time, however unheralded they might have been, will live on long after them. That is, in atheist terms, a most considerable means of achieving not-death.

Atheism through Reason

The intellect must be engaged to undertake atheism. The intellect responds to logical thinking, which is thinking by rules, and avoiding fallacious thinking. We discussed one such fallacy earlier in this book: *argumentum ex silentio* – assuming the absence of evidence is proof of the existence of that which is absent. There are other such fallacies:

- The straw man argument (misrepresenting the position of the opponent)
- *Argumentum ad hominem* (attacking the opponent rather than the argument at hand)
- Begging the question (using the conclusion sought as proof of an argument)
- The false dilemma (assuming only two answers are possible, as in the assumption there are only two moral forces at work in the world - good and evil)
- Raising the bar (demanding more evidence after an argument has been proved)
- *Onus probandi* – shifting the burden of proof on to the opponent (very commonly used by believers – "Prove that God doesn't exist!")
- *Argumentum ab auctoritate* (appealing to a supposed authority, such as the Holy Bible)
- *Argumentum ad consequentiam* (argument by consequence, such as "You will go to hell if you do not believe in God")
- *Argumentum ad antiquitatem* (appealing to tradition, as in "Our family has always believed in God")
- *Argumentum ad populum* (appealing to popularity – "Everyone else believes in God, why don't you?").

This is not to say that believers in God must abandon their intellect. Some of the greatest thinkers of the Western world have been theologians such as Thomas Aquinas and St. Augustine. But to defend belief in God and religion, it is necessary ultimately to resort to non-logical arguments, to make that *leap of faith* which says "If you believe this with a strong enough faith, it will be so." To go beyond defending the faith, and proceed to attacking the non-believer, it is inevitable that some sort of fallacious and illogical argument, such as any of those listed above, must be employed. And if this does not work, ultimately the believer can resort to emotional appeals, since belief in God ultimately arises in the unconscious mind, and tapping into that unconscious, emotional impulse is often the most productive way of converting someone into believing in God.

You can see, therefore, what the atheist is up against. The atheist cannot use emotional appeals, and certainly not fallacious logic, to make their case. Logical thinking is careful thinking, and involves the search for truth or falsity in statements, including those which are made constantly and are simply assumed to be true. Probably no statement is made more frequently, and assumed to be true, than the statement that God exists. The ubiquity of this belief is such that it would not even occur to most people to question its truth, much less overcome a lifetime of religious thought and practice which is based ultimately on the assumption the statement is true.

The first step to developing an atheistic way of thinking is to ask the simple question: is the statement that God exists true or false? Asking that question is a significant breakthrough, because it represents an unwelcomed challenge to the authority of all Regents and scriptural sources which assert the statement is true. It is the sort of question that arises in the mind of a believer first as a niggling doubt. The believer must then overcome all the built-in mental defenses of faith in order to take yet another step, and that is to investigate the question seriously. Merely reading a book such as this, or any other of the many books about atheism, is one of those next steps, because it means the believer has overcome the defensive assumption that books about atheism are inherently the works of Satan and automatically condemn the reader to damnation in hell.

Those who pursue the question of God's existence by using reason and logic must ultimately conclude that God does not exist in any material sense, but only as a compelling mental illusion. Some philosophers and theologians have tried mightily to reconcile God-belief (faith) with reason. The French philosopher Blaise Pascal said:

> **If we submit everything to reason our religion will be left with nothing mysterious or supernatural. If we offend the principles of reason our religion will be absurd and ridiculous . . There are two equally dangerous extremes: to exclude reason, to admit nothing but reason.**
>
> **-Blaise Pascal**

Pascal made significant contributions to the study and science of logic, and he tried to reconcile faith and reason by proving God's existence through logical reasoning. The result is famously known as "Pascal's Wager," which is flawed on several levels, not the least of which is that Pascal assumed that the bettor places the wager on the existence of the right God, which in Pascal's terms was the Christian God. Anyone choosing the wrong God would be condemned to hell, even though they had satisfied the terms of the bet (belief in God) which supposedly would lead to a winning solution.

In some cases, philosophers had doubts about God's existence, but did not wish to risk writing about God from an atheistic perspective. The German philosopher Georg Wilhelm Friedrich Hegel's philosophical masterpiece, *The Phenomenology of Spirit* (*Phänomenologie des Geistes* - 1807), originally was written with the emphasis on the meaning of *Geist* as Mind, not Spirit (the word can mean either in German). Hegel explored consciousness in this book, and investigated how consciousness proceeds to self-awareness, reason, knowledge, and ultimately the development of man-made belief in God and religion. Unfortunately, one of Hegel's colleagues was at the time fired from his university position for having written a book espousing atheism. Hegel took the lesson seriously, and altered his book to emphasize that *Geist* meant Spirit, or God by implication, and that Spirit existed independent of the Mind. Hence, God existed independently of man, and religion was a proper structure in which to explore, understand, and then worship God.

Most other philosophers and theologians have simply written in favor of faith as proof of God's existence, over anything reason can offer.

> **Reason is a whore, the greatest enemy that faith has; it never comes to the aid of spiritual things, but more frequently than not struggles against the divine Word, treating with contempt all that emanates from God...Faith must trample under foot all reason, sense, and understanding.**
>
> **-Martin Luther**

> **Reason is in fact the path to faith, and faith takes over when reason can say no more.**
>
> **-Thomas Merton**

> **The search for reason ends at the known; on the immense expanse beyond it, only the sense of the ineffable can glide. It alone knows the route to that which is remote from experience and understanding.**
>
> **-Rabbi Abraham Joshua Heschel**

> **Atheists are my brothers and sisters of a different faith, and every word they speak speaks of faith. Like me, they go as far as the legs of reason will carry them -- and then they leap.**
>
> **-Yann Martel**

> **Intellect takes us along in the battle of life to a certain limit, but at the crucial moment it fails us. Faith transcends reason. It is when the horizon is the darkest and human reason is beaten down to the ground that faith shines brightest and comes to our rescue.**
>
> **-Mahatma Gandhi**

Many other defenders of belief in God have written in a similar manner, and their argument ultimately boils down to the observation that reason can take mankind only so far. Reason is, indeed, constrained by Nature, which is why it is so closely associated with the natural sciences. We apprehend Nature through our five senses, and through whatever scientific tools we devise to explore Nature beyond our senses, such as the x-ray machine, the telescope, the seismograph, and the particle accelerator. In some way or another, all of these tools are *extensions* of our senses, which allow us to see further or in more detail. The particle accelerator, for example, allows us to infer the existence of the most fundamental building blocks of Nature, such as subatomic particles like quarks and muons. While we cannot see these particles, we can at some point be assured of their material existence; they are not entirely invisible to us.

God, in the minds of the most sophisticated theologians, is entirely invisible, and cannot be discerned through a device such as a particle accelerator. To accept his existence, theologians say we must go beyond physics, into *metaphysics,* which in its purest sense posits that mind is existence and the transcendental essence of the universe. Theologians and metaphysicians argue that when we come to the shoreline where the natural sciences end, we must "leap" or "sail across the waters" or "glide" unto the ineffable – that which is of such enormity of consequence that we cannot comprehend it. This is how we come to know God, in the opinion of great philosophers and theologians. This is the point where "faith shines brightest and comes to our rescue," according to Gandhi.

There is something grand, glorious, and seductive about this way of thinking. Man is indeed capable of perceiving the limitless. We can approach the ineffable, knowing we shall never truly reach it. Our imagination can create the most wonderful ways of describing this land beyond the oceans, this realm of the gods – Elysium, Valhalla, Heaven, the Pure Land, Oneness with Brahman. That they are one and the same place, and that they derive their creation from our real, infant experiences in heaven, escapes the mind of most philosophers and theologians, because they are still in thrall to their unconscious desire for God and heaven, as is most of mankind. They have not freed themselves from the urgings of the unconscious mind.

It is only through atheism that this can be accomplished. And when it is accomplished, what happens to the Pure Land, to Oneness, to Elysium, and to any other manifestations of heaven? They are still there, in the recesses of our mind, as is the longing for heaven. They can still be acted upon by those who do not believe in God. This is how mankind makes advances in science, and this is how the shoreline at which Nature ends is constantly being expanded and pushed forward, reducing the size of the ocean of ignorance and doubt that lies beyond. As was said by two noted scientists:

> **For it is the chief characteristic of the religion of science that it works...**
>
> **-Isaac Asimov**

> **... genuine religiosity does not lie through the fear of life, and the *fear of death*, and blind *faith*, but through striving after rational knowledge.**
>
> **-Albert Einstein**

Note that Asimov and Einstein refer to science as a religion, and rational knowledge as the pursuit of genuine religiosity. It is this tendency of scientists to treat the discipline of science as a form of religion that has drawn the interest of many theologians. An example is the earlier quote above, by Yann Martel, who wrote *The Life of Pi*, who said that "atheists are my brothers and sisters in faith." It is true that atheists and theists, and all other seekers of not-death, are following similar paths. It is not true, however, that the path of atheism is one of faith. The path of atheism is through the application of logic and reason, and faith is a defense against the use of reason. It is a contradiction to believe that atheists are somehow no different from religious people, and that they have their own religion called atheism. This is not only a fundamental misunderstanding of atheism, it is a condescending statement that is designed by theists to minimize the basic differences between atheism and faith.

Atheism requires an objective look at oneself, a stepping out of one's personality to observe from outside and from a distance those walls of faith that have been erected to defend God-belief and religion. Only then can doubts regarding God and religion be pursued with seriousness. But it is by no means an impossible task. It helps considerably to have empathy for others, because empathy is the act of stepping out of one's personality and conceiving what it must be like to be someone else. Empathy is an aspect of altruism, and as we've seen earlier in this book, altruism is a form of goodness.

The concepts of altruism and goodness give us a real clue about what atheism, in all its different pathways and end points, is all about. If you wonder how everyday sacrifices for the benefit of others can truly matter, consider this list of eminences: Aristotle, Bertrand Russell, Eleanor Roosevelt, Alexander Hamilton, Nelson Mandela, John Keats, Malcolm X, J.R.R. Tolkien, Racine, Somerset Maugham, Ingrid Bergman, Leo Tolstoy, Johann Sebastian Bach, Johannes Kepler, Jean-Jacques Rousseau, and George Washington Carver. All these people had one thing in common: they were all orphans, raised by relatives, friends of their deceased parents, or in an orphanage. Who were all the individuals who housed, fed, clothed, and educated these people? We'll never know all of their names. We'll never know if they were religious, indifferent to religion, or fervent doubters in the existence of God. But it really doesn't matter. We all owe them a debt of gratitude for their selflessness – for saving a life of great destiny, from possible destitution and lack of opportunity.

The atheist is aware, to a degree that religious people often are not, that human progress, and the advance of civilization, depend not only on the greatness of certain men and women of renown, but on the sacrifices of ordinary people which create the opportunities, environment and community where great people may flourish. The atheist focuses their attention, and their ambitions for not-death, on the human timeline, on that long arc of human history which is the product of secular advancement even more so than religious achievement. The human timeline is stabilized by religion, which offers society a relatively fixed or slowly changing dogma governing human behavior. The human timeline is disrupted by secularism, usually in a material way, and more often than not for the good, as long as the secular change is undergirded by an impetus for the betterment of mankind. Atheists, with their behavioral predilection toward empathy, and their application of altruism on the individual and community level, help keep secular change on a positive track.

The Pathways to Atheism

There are many pathways to atheism, all of which tend to move in the opposite direction of religious belief, especially within the Abrahamic traditions. The Abrahamic religions – Judaism, Christianity, and Islam – all have in common certain shared scriptures and religious figures (such as Abraham himself). They are monotheistic religions, for the most part (Christianity shares three aspects of God in one Godhead). They rely on faith to buttress religious belief. Atheism has no scripture, and is the opposite of faith. In between these two disciplines, both of which are legitimate roads to achieving not-death, are the Eastern religions and related traditions, such as Buddhism, Shintoism, Confucianism, Taoism, and Hinduism. These religions are often described as "Righteous Path" traditions, where the believer is encouraged and expected to follow a path of personal improvement, in a tradition which is less hierarchical in structure and less focused on sin and a blood sacrifice as a means to providing mankind a form of redemption.

Atheism requires that the individual must first look at themselves from a distance, with objectivity. Once the atheist begins to doubt the existence of God, a rush of questions ensues. If my belief in God and religion has all along been incorrect, why did I believe these things in the first place? If these beliefs were given to me by my parents or caregivers, as well as the religious community in which I was engaged, and if they were first given to me as a child, what were the motivations of those who did this? If I abandon my beliefs, what are the consequences for the relationships I have with my family and friends? To whom do I tell about my atheism? If I lose important members of my support group as a consequence of telling them, how will I survive in this new world? To what degree am I angry at the situation I have been placed in by my family and friends, and how do I deal with this anger long term?

There are many useful books written by atheists about their journey away from God-belief and religion, and some of the better ones are referenced in the bibliography at the end of this book. Because these journeys are always personal, we will only discuss here the different paths people can take to achieve atheism.

The general assumption is that the road to atheism is linear. We can presumably stop along the way, and strengthen our disbelief, but at the end lies true atheism, which some define as complete disbelief in God or gods, without any doubt whatever. The pathway, therefore, supposedly starts at doubt, and ends in atheism. This is a useful way to think of atheism, as a spectrum measured by intensity of disbelief. But it is a limiting metaphor.

It is better to think of atheism as consisting of multiple paths, no one of which is more accurate or worthy than any other, and none of which needs to be a terminal point where we can abide for the rest of our life.

Here is a set of different pathways one can take to achieve atheism:

Doubt

Doubt regarding the existence of God and the value of religion is the first step any of us must take in order to explore atheism. It is not necessary for any of us to go beyond this point. A good many practicing believers have doubts about God. Sometimes these are temporary doubts, when it appears that God has inflicted too much misfortune on us, against which we are no longer able to cope. Sometimes a single misfortune, like the accidental death of a child, is enough to create a lifetime of doubt. None of this means any of us in this situation needs to abandon religion, if we find that the benefits of religious Community and Agency are too important to give up, and as long as we fully comprehend the costs of God-belief and the potential abuses which can be perpetrated by religion. These same abuses can be committed by those outside of religion, including atheists, but religion disguises or hides such abuses through the gloss of moral superiority which it projects. Because of this, many people who doubt God's existence stay for a time with their religion and religious community, but eventually take other steps that lead to a rejection of the idea of God. This next step often is a painful process, because the atheist who goes beyond doubt also must confront the falsity of their previous beliefs. A sense of betrayal sets in, and often develops into anger at the family and friends who nurtured these false beliefs.

Rejection and Anger

It is very frequently the case that those who leave religion and abandon belief in God find themselves deeply angry. They are hurt from the realization that they have been deceived to a degree, by their parents, family, friends, and community. The family and community they have left behind do not, of course, sense that there was any deception, since they are all of the conviction that they are in possession of truth regarding God and the afterlife. This enforces the sense of estrangement for the atheist, who has difficulty understanding how others could accept fantasy thinking. Part of the anger comes from the fact that overcoming all the defenses of faith in God is a strenuous, difficult, and emotionally painful act. It is no wonder, then, that atheists are often seen by others as angry, because many are, and many retain some of their original anger as they negotiate their new world of atheism, is which they are ostracized, ridiculed, and sometimes endangered by society in general. It is easy in such circumstances for an atheist to blame God, or at least what faint image the atheist may have of God:

> **He was an embittered atheist (the sort of atheist who does not so much disbelieve in God as personally dislike Him), and took a sort of pleasure in thinking that human affairs would never improve.**
> — George Orwell

Additionally, to the extent religion is given a privileged position by the state, all the problems of abuse of power and organizational malfeasance which can plague a religious organization now are amplified many times over, as government can either abide such abuses or prosper from them. The ferocity with which the Catholic Church was treated during the French Revolution was the direct result of the privileged position it was given by the monarchy and the state, and resentment of the Church persisted well into the 19th century. This is exemplified by the following quote:

> **Civilization will not attain to its perfection until the last stone from the last church falls on the last priest.**
>
> -Émile Zola

Most atheists understand that anger is a path with an unhealthy termination; no one wants to be perpetually angry and at odds with the world. Most atheists therefore seek out other pathways of atheism, some of which use this anger for productive purposes.

Humanism

The most frequently explored pathway of atheism is humanism, sometimes called Secular Humanism, which represents the conviction that each of us has unexplored potential to improve the human condition. Nor is humanism, despite its title, restricted simply to the betterment of mankind as its ultimate purpose. Efforts to improve the conditions of all living things is an important part of humanism, just as it is with some religions such as Jainism. Humanism offers such rich avenues for exploration and service that it is impossible here to summarize them all. We will touch more thoroughly on humanism in our final comments of this book. For now, it simply needs to be said that there is nothing that religion can accomplish through its Agency function that cannot be accomplished through humanism.

Agnosticism

Agnosticism is usually described as the belief that it is impossible to prove one way or the other that God exists. In this strict definition, agnostics should have an affinity with human-ism, since working to improve the world while we are in it is the best option available if it is impossible to prove whether God exists. Many agnostics do indeed identify themselves as humanists and try to live their lives in the most "humane" way possible. The more general interpretation of agnosticism is that agnostics do not know whether God exists, and this is a subtle difference from the first definition. To not know whether God exists is to leave open the possibility that he does indeed exist. By this definition, the agnostic is simply a doubter regarding God's existence. Bertrand Russell expressed the difference this way:

> **As a philosopher, if I were speaking to a purely philosophic audience I should say that I ought to describe myself as an Agnostic, because I do not think that there is a conclusive argument by which one can prove that there is not a God. On the other hand, if I am to convey the right impression to the ordinary man in the street I think**

**that I ought to say that I am an Atheist, because, when I say that I cannot prove that
there is not a God, I ought to add equally that I cannot prove that there are not the
Homeric gods.**

-Bertrand Russell

By this standard if someone is a doubter, they are an atheist overall, and given our earlier comments in this book on faith, that would be because they have overcome the defenses used to sustain faith in God. On this point, many agnostics might disagree, since many of them do not want to be "tainted" with the label of being an atheist. The word has such pejorative resonance, especially in America, that it is common to inflate the difference between atheism and agnosticism, to avoid the accusation of being an atheist. Also, within the atheist community, those who are atheists proper sometimes denigrate agnostics as wishy-washy temporizers. This leads to the sense that there is a greater difference between atheists proper and agnostics, than between atheists proper and religious believers. This was the subject of a jest by an American film director.

**I did not marry the first girl that I fell in love with, because there was a tremendous
religious conflict, at the time. She was an atheist, and I was an agnostic.**

- Woody Allen

Activism

Many atheists tend to keep their atheistic beliefs to themselves, and as we shall see in the next section, there are very good societal reasons for doing so. In some cases and countries it is a life and death matter not to reveal one's atheism. Other atheists rebel against these external constraints, and at least in countries where it is relatively safe to do so, they proudly proclaim their atheism. Atheists can advocate for their disbelief in God among their family and friends, and while this is not often the case, advocacy can turn into proselytizing. Where advocacy for atheism really makes a difference, however, is at the societal level. The most popular books about atheism in the past twenty years have been advocacy books, which while promoting atheism, also attack and frequently ridicule religion. These books are examples of activism being combined with anger, or better said, activism used to channel anger into productive purposes. There are questions in the atheist community as to whether this is the most productive approach to promoting atheism, and some of these questions are motivated by the fact that many people do not have the personality for public activism. But there should be no doubt that such books, angry though they may seem, are essential, because as we shall see in the following section, atheism is often a lonely journey in a society which defaults to God-belief and religion. Moreover, many parts of society are actively antipathetic to atheism and atheists, and the only proper response to such activism is public advocacy on behalf of atheism and atheists.

Another important aspect of activism is protecting the rights of atheists and preventing governmental or government-funded institutions from promoting religion. This struggle is most prominent and organized in the United States, where organizations have been set up to use legal means to prevent school administrators from engaging in public prayer, government offices from displaying Christmas crèches, and churches from using government means to preach and proselytize. This work is in direct response to deliberate, well-funded, and highly organized efforts by conservative Christian denominations over the past forty years to "move Christ back into the public square." The struggle in the United States between church and state is particularly fierce, especially since Christians are not used to getting pushback from secular organizations, and since public participation in religion is on a sharp decline, especially among younger people.

Despite these efforts, it should be noted that other than in Europe, atheism in most countries is unwelcome by society at large, including governments, and in many places it is actively if not viciously opposed. The work of activists to turn this situation around is very limited and often quite dangerous in places other than the industrialized West.

Community

Some atheists have realized that they miss the organized community aspect of the churches, synagogues or other religious institutions they have left behind. "Sunday services" have been created in the U.K., the U.S., Australia, and elsewhere which provide a way for atheists to meet weekly for singing, lectures, food, and other forms of conviviality. Humanist societies are also actively involved in setting up monthly meetings, creating charity drives, inviting in speakers and authors, and promoting their organization publicly. The idea that atheists need a community is quite new and is not universally accepted among atheists. Many atheists feel that their process of moving away from God-belief involved also moving away from any organized or ritualized gathering, religious or otherwise.

Still, while there is not universal agreement within the atheist community that it needs Sunday services – that it is even a community by any definition – atheists are not immune to mankind's universal desire to be part of something bigger, some tribe or group of like-minded people. This is especially so considering that in most societies, atheists are still publicly shunned and often find comfort in the company of other atheists who understand what it means to be purposely isolated by a society which finds their disbelief discomfiting.

Atheism Proper

This form of atheism represents the traditional, public impression of atheism, and consists of individuals who reject outright the existence of God in any form. As described above, the denial of the existence of God is the consequence of the process of atheism, but our definition here is a minority opinion. Most people think "Atheism = Disbelief in God." One purpose in this book of challenging this traditional definition of atheism is that it is often in the public mind converted into something pejorative, such as "Atheism = Rejection of God," or "Atheists = Immoral Godless Heathens," or "Atheism = the Opposite

WHO CUT GOD'S HAIR

of Religion." These negative impressions lose sight of the many varieties of disbelief that constitute atheism: from uncertain (wherein some people still practice their religion) to very strongly held disbelief. They also lose sight of the arduous process and struggle many people go through to shed their religious beliefs and enter into doubt as to God's existence.

Consequently, we will stay with our definition of atheism as a process of shedding God-belief. Atheism proper, therefore, we define as the complete shedding of the walls of faith, leading to the strongly held conviction that there is no God in external or any other form, and the universe was not created by a Supreme Being or any other form of deity.

The Lonely Road

American presidents commonly invoke God and their faith as the greatest bulwark in support of the burden they must carry as chief executive:

> It is hard to see how a great man can be an atheist. Without the sustaining influence of faith in a divine power we could have little faith in ourselves. We need to feel that behind us is intelligence and love. Doubters do not achieve; skeptics do not contribute; cynics do not create. Faith is the great motive power, and no man realizes his full possibilities unless he has the deep conviction that life is eternally important, and that his work, well done, is a part of an unending plan.
>
> — Calvin Coolidge

If you have grown up all your life listening to your country's leaders demean or mock atheism, and extol the virtues of their faith, you get the strong impression that the last thing you would want to do to get ahead in life is become an atheist. It takes tremendous effort of will to break away from religion and faith – to reject faith in favor of atheism. Even those who explore atheism in their childhood tell of feeling estranged from other children and their family, depending on how religious their family was. The real effort, however, starts for most people in their teenage years and beyond. How does one abandon the beliefs and practices of a lifetime?

It starts as an internal struggle, a battle of intellect and reason, fighting the emotional comfort we tend to derive from belief in God as an ally who will help us avoid death. The comfort arises internally from our unconscious desire for a God to replace the one lost as a child. Our unconscious mind cries out for security, a lack of want, a reduction in stress, a suffusion of love, and an elimination of death – all of which we feel is our birthright. We don't dwell upon these thoughts consciously. Death in particular we banish from our thoughts the minute some external event reminds us of it. Yet the anxiety of providing for ourselves and others; of dealing with the imperfect love which is the highest form of love any human can offer us; of stressing incessantly over our health; and of reminding ourselves as often as necessary that death is a far-off eventuality – all of this stirs up our unconscious desire to be done with this world in the form in which it exists. All of this cries out for a father figure to save us, and if God the Father cannot appear to us now in our real life, the knowledge that he once existed in our real world, and that we will join him in the afterlife, is enough of an analgesic that we are easily satisfied.

The internal battle, then, consists of reminding ourselves consciously, in conversations we have with ourselves and with others, that God the Father is nothing more than Illusory God, and if we accept him in our life we gain fortitude in the face of the truth of our existence, but we lose as much in terms of our ability to avoid the impositions of religion. This includes bearing the burden of sin and hellfire, and losing the capacity to think for ourselves about morality and many other important matters.

The battle is ultimately decided on our ability to face directly the truth of our mortality. The human species has gone through enormous efforts to deny the simple fact that death is inevitable, it is part of the natural order, it is final, and it leads to the same oblivion we inhabited before we were born. The key word in this list is *natural*. The atheist comes to accept that humans are a part of a natural order which always has and always will result in death for all animate things, and ultimately even the inanimate home on which we thrive, planet earth. An atheist embraces the natural order of things, and comes to terms consciously with their own inevitable death and oblivion. Is it any wonder, therefore, that those whose life work is involved directly with the natural sciences, should find it easier to accept that Nature, not God, defines whence we have come and to where we are going?

> **Nature is what we know. We do not know the gods of religions. And nature is not kind, or merciful, or loving. If God made me — the fabled God of the three qualities of which I spoke: mercy, kindness, love — He also made the fish I catch and eat. And where do His mercy, kindness, and love for that fish come in? No; nature made us — nature did it all — not the gods of religion.**
>
> **— Thomas A. Edison**

God-belief and religion demand the opposite: a rejection of Nature in favor of super-nature, or the supernatural; a belief that God gave mankind command over Nature; and acceptance of an elaborate theology, by no means unique to Christianity, which states that command over Nature ultimately implies command over death itself. Atheists, for example, accept evolution as part of the natural order; on the other hand, some highly vocal God-believers (and not only Christians) reject evolution and embrace instead a mythology that insists that God created all creatures all at once, forevermore.

Atheism takes an altogether different form of courage than theism, which requires only the courage to suspend disbelief in favor of ancient mythologies. The courage required of the atheist includes the ability to accept the scorn of the religious world.

The World Fears the Atheist

All humans fear death, and all humans repress thoughts of death. Those who believe in God go a significant step beyond this, by smothering their fear of death with a blanket of denial, crafted as mythology, sold as religion, and reinforced through ritual. But in so doing, the believer picks up a new form of anxiety, beyond all those fears imposed by religion. This new form of anxiety is a conscious suspicion that something isn't quite right with God-belief, and religion in particular. Something about God-belief is illogical, especially the requirement that the believer adopt the behaviors of an infant – bowing and kneeling and otherwise crawling along the ground in supplication to an invisible entity. Worse, deference must be paid to God's Regents, who will always have the faults of a human. Worst of all, the believer carries a lifetime fear of himself or herself, and the utter worthlessness of their internal character – the ugly sinful aspect of their being – that must disgust a being as pure as God.

These suspicions, illogicalities, and fears are among the costs of God-belief and religion, and they are borne for a lifetime. But they are not generally able to compete with the much greater fear of death, and so the believer accepts them as part of the human condition. When in the presence of an atheist, however, the believer is troubled by the suspicions, illogicalities, and fears that lurk in their own conscious mind. The atheist's mere existence reminds the believer of the costs of their own belief in God, and the believer deeply resents this reminder.

This resentment expresses itself in several ways. We have already talked about the element of psychological projection that the believer imposes on the atheist. The atheist is accused of having no moral compass, when in truth most atheists think carefully about the morally right thing to do, and are more versed in matters of morals and ethics than many believers. The accusation of immorality is nothing more than a reflection of the believer's moral crisis: what if there was no religion? Who would teach them morals? What horrible crimes would they commit? How could they live without a Regent scolding them for their sinful behavior, and without a God to punish them in this life or the next? These questions are motivated by the fact that the believer is not allowed to develop their own moral or ethical system – everything is spoon fed them in little bites that have gone through the filters of theological orthodoxy and cultural conditioning.

The atheist, merely by their presence, reminds the believer that there are people in this world who do not carry such burdens, questions, and fears. The atheist is always therefore an irksome presence to the believer, precisely because the believer does not accept that anyone can live without these burdens, questions, and fears, and precisely because the believer is deeply jealous of the atheist for enjoying what appears to be an unimaginable psychological luxury.

Some Regents respond much more violently to atheists than an average believer might, because the Regent has so much more at stake. The Regent is responsible for the correctness of the theology being presented to his or her congregation. The theology has to be in accord with what is generally being presented by the religion in question, and specifically with what the Regent or his or her overseers (if there are any) will allow. No member of the congregation may challenge the Regent in the substance of that theology, and even in the give-and-take questioning that may accompany an informal study of the theology in something like a Bible study program, the Regent ultimately gives the group the correct theology, which they must accept rather than question. So strong is the authority of a Regent within their own congregation, that many Regents would be surprised, if not horrified, that someone in the congregation was seriously challenging the Regent on a theological question.

There is a doubling of that horror when a Regent is confronted by an atheist, either as a party known to many of the congregation, or as someone completely unknown to the congregation. The atheist will be shunned, and their teachings declared anathema. If the atheist is a member of the congregation, they can easily be expelled. If the atheist is a young adult under the control of a family within the congregation, Regents have been known to interfere in the family structure by ordering the parents or guardian to shun the atheist and prevent any further communication with the other members of the family.

Shunning is, as we've seen earlier in this book, a passive form of violence against the victim. Its only possible redeeming quality – and this is small comfort - is that shunning is not active violence inflicted on the non-believer. But the world is hardly lacking in such violence. Atheists are feared and despised worldwide, because they threaten the power of Regents. Just in the past few years, atheists have been thrown off three-story buildings to their death (Syria), threatened with execution by beheading (Saudi Arabia), whipped with canes and lashes (Malaysia), stoned to death (Syria), and strangled to death as a crane lifted them slowly off the ground by a rope tied around their necks (Iran). In Saudi Arabia it is now a capital crime merely to espouse atheism. In North Carolina a few years ago, a man entered the Sunday service of a Unitarian Universalist church and shot and killed two people, because the church tolerated atheists in their congregation.

Given the ugly reality of how a world dominated by religion can sometimes treat atheists, no one should wonder why some atheists are angry, and no one should doubt that it takes courage to explore and achieve atheism.

Confronting Mortality

The highest challenge of the process of discovery that is atheism is confronting death as an inevitable aspect of the human condition. Atheists abandon all of the psychological deceits that protect us from fear of death, such as belief in the afterlife and the promise of eternal presence with God. Instead, atheists confront their own mortality head-on, with the full understanding as well that nothing follows death. The act of dying may itself be painful and to be feared, but death itself, in the atheist view, results in oblivion, or the condition we were all in before we were born.

Death is, therefore, eternal sleep, without the unconscious mind working in the background to provide dreams. As eternal sleep, death has something in common with the Christian wish for *Requiescat in Pace*, or Rest in Peace. *Requiescat in Pace* is one of the earliest implorations of Christians, and was more than likely borrowed from Judaism. It represents a desire for rest after the tribulations of life, as distinct from the usual expressions of hope for an eternal life in Heaven.

Isn't this a contradiction? How can someone have eternal life in heaven with God, all of their relatives, friends, and others who were saved from damnation, yet at the same time be sleeping? Christian theologians have finessed this contradiction by asserting that the rest believers seek is not really eternal, but lasts only until the "Last Trumpet Sounds," when Jesus will return at the End of Time to judge all humans, living or dead. But isn't this a contradiction as well? Most people are encouraged by their religion to believe their relatives and friends who have died, if they were good, religious people, are now in heaven with God. Why else would people send them intercessory prayers, asking them to help the rest of us alive here on earth?

The real contradiction is that we want both things: heaven, and eternal sleep. With eternal sleep, we know what we are getting, if it is truly the slumber of the unconscious, with no conscious awareness of dreaming (no nightmares to wake us up, for example). Eternal slumber, which is equivalent to our condition before we were born, is nothing to be feared, and in some respects tantalizing compared to heaven. How can anyone tolerate, for example, an eternity of bliss with God in heaven, being required to praise him constantly, unrelentingly, for all time? Wouldn't unremitting ecstasy become the equivalent of unremitting agony? An eternity of anything is something to be feared, if only from the boredom of it all.

St. Paul said, "In a moment, in a twinkling of an eye, at the last trump....we shall be changed." Perhaps he meant that at the Last Trumpet, the elect – the saved – the saints granted eternal life in heaven – shall be changed into angelic form. Our worldly condition, and the particulars of our human body, will disappear, as we are conjoined with God in perpetual bliss. Boredom won't come into the matter. Perhaps this is what he meant, but it is still unsatisfying. We cannot identify with the angels. Our mind can only grasp the human condition, and many people at the end of life are weary of the burden, especially the pains of old age. We should not be surprised, therefore, as we "shuffle off this mortal coil," or burden, to use Hamlet's phrase, that we would welcome eternal sleep, the slumber of the just – and be satisfied with that instead of heaven.

If eternal sleep is benignly attractive, why would so many people still long for heaven? Because it was once real in our lives, and is something on an unconscious level we can relate to. More important, heaven is a compelling answer to the problem of the truth of our mortality. Unless we are in great physical pain or mental anguish, we unquestioningly run away from death. We repress even the thought of it, and for most of our life, our instinct is to cling to this mortal existence. Ultimately, our existence on earth is all we have, it is all we have known, and as it is the most important thing to us, it is the one thing we wish to prolong *forever*. Hence, we invent heaven as a means of cheating death, and just as importantly, extending our existence on earth, with the company of those family and friends we knew *and liked* while we were alive. Aunt Hortense, for example, would not be allowed in if she was always nasty to us here on earth. And we certainly would not want to go to a heaven where we knew no one, where everyone there spoke a completely different language, and where we would be perpetually lonely.

For those who have achieved atheism, and who have rid themselves of the illusion that there is a heaven, the idea of eternal sleep is intellectually satisfying, but not necessarily emotionally comforting. The prospect of eternal sleep does not smother the fear of death, because eternal sleep may be not-death, but it is not life, and it is the extension of our life for as long as possible that we desire.

How do atheists cope with this desire for not-death, in a form that is more than just eternal sleep? This is the subject of our final chapter in this journey of discovery.

CONCLUSION

CHAPTER NINETEEN

What is it We Seek?

"I don't say he's a great man. Willie Loman never made a lot of money. His name was never in the paper.

He's not the finest character that ever lived. But he's a human being, and a terrible thing is happening to him. So attention must be paid. He's not to be allowed to fall in his grave like an old dog. Attention, attention must finally be paid to such a person."

-Arthur Miller

Attention Must be Paid

The despair of Willy Loman, the central character in Arthur Miller's *Death of a Salesman*, is that he devoted his entire adult life to a corporation which, when he had ceased to be useful, fired him. There was no thank you, no pension, no recognition of any sort that his service of a lifetime had any meaning. In his retirement he developed a passion for planting seeds, and watching vegetables grow; seeking, in other words, anything that would be left behind on this earth that could be said to be his contribution.

It is Willy's wife Linda who demands of her sons that "attention must be paid" to a man like Willy Loman. She is the only one in the play who senses that her husband is thinking of committing suicide, and this is more than anything the "terrible thing" that is happening to him. The despair at being a failure in life, the cruel rejection he has received from his employer, his disappointment in his sons – all of these tribulations weigh on Willy Loman enormously and prove to be too much to handle. In the final scene at his funeral service, Willy's family comes to different conclusions about his life and what it meant to them.

Great art should never provide definitive answers to life's most important questions. Instead, the artist should pose the questions that matter to us most, and then invite the audience to come up with its own answers. *Death of a Salesman* continues to challenge audiences nearly seventy years after its debut. Was the American dream of wealth and popularity that Willy chased all his adult life a false idol? Was Willy right to place so much hope in his sons, expecting them to be the business success he was not? Of whom was his wife speaking when she said "attention must be paid" to someone who was seemingly abandoned by the world and about to be dumped into his grave "like an old dog?" Should his employers have paid more attention to him? His family and friends?

What did the world owe Willy Loman? Willy is not an especially virtuous character, and many people in the audience turn against him when it is revealed he's been cheating on his wife Linda. In that respect, the world owes him nothing. While it appears at first that the tragedy of Willy Loman is the way the world has abused him and then forgotten him, the real tragedy of Willy Loman is of his own doing. By the time he is in despair for his life, his attempt at achieving not-death by planting vegetable seeds is futile. There is no one around him who has any interest in tending to his vegetable garden in his absence, just as no one at his funeral service can find the right words that would reflect on anything meaningful he ever did in his life.

Nor is it evident that Willy Loman would have fared better in life had he been religious. A book could have just as well been written not about a selfish salesman, but about a religious figure who also lacks any positive meaning in his life. In fact, one was – *Elmer Gantry*, by Sinclair Lewis, features an evangelical minister who is a complete hypocrite, gambling and lying and chasing after women during the week, and preaching the gospel on Sundays. This is a model of a preacher all too familiar to modern audiences if they pay any attention to the falls from grace which plague so many American televangelists.

Attention must be paid, but not so much by the world at large, as by you and me. We, the audience, must take our own lessons from the fictional characters of Willy Loman and Elmer Gantry, or if we prefer, from the real persons who all-too-often dominate our news feeds and disappoint us. The billionaire businessman who makes a career out of cheating clients and workers; the politician with big promises who obtains office by stimulating the fears, bigotry, and ignorance of the voters; the religious leader who preaches the Prosperity Gospel as a rationalization for his own obsession with the trappings of wealth – all of these people obtain fame, fortune, power, and respect. Their hero project is played out on a national or global stage, and fascinates millions of people. And yet there is nothing truly heroic about them. For all their grandiosity, they are failures regarding what is important in life, and that is the necessity to have a positive impact on those they meet during their journey here on earth.

For what will the world remember Bill Gates? As the founder of a major software company, or isn't it more likely he will be remembered as a man who, in partnership with his wife Melinda, used his fortune to help eradicate malaria from this world? Is there a modern politician whom historians will respect more than Nelson Mandela, a man who spent most of his adult life in jail, and who, when he was elevated to the position of president of South Africa, insisted that the black man and the white man had to work as equals, and that neither was superior to the other? For what will the world – meaning those people who knew you personally – remember *you* when you have gone? Your career? Your wealth? Your position in society? Or will it be those qualities you displayed in life that favorably impacted others – qualities of altruism, quiet kindness, patience, generosity, and that disposition of grace that can be read as spiritualism, or as a secular confidence in the humanity of others?

What about those who are unsure of their contributions? Mother Teresa throughout her life confided to a priest that she had a "darkness" in her soul that prevented her from reaching God through her prayers. She often doubted he even existed, and eventually gave up praying altogether.

In the 1930's, Father Pierre Teilhard de Chardin wrote what is now considered a theological masterpiece, *The Phenomenon of Man*, in which he argued mankind would continue to evolve until our species would ultimately merge with the divine. His exposition of his theory was scientific in nature, based on his work as a paleontologist and geologist, and his deep belief in Darwin's theories of evolution. It was his argument that Christ's Incarnation was a critical event in the history of mankind, directing us firmly on a path to union with God. After Teilhard's death, it was discovered in his private letters that while he was sure he was correct that "the World" would continue to evolve, and man was destined to achieve union with God, he had fundamental doubts that Jesus Christ had anything to do with it:

> **"If by consequence of some internal upheaval, I came to lose successively my faith in Christ, my faith in a personal God, my faith in the Spirit, it seems to me that I should continue to believe in the World. The World (the value, the infallibility and the goodness of the World), such in the final analysis is the first and the only thing in which I believe. It is by this faith that I live, and it is to this faith, I feel, that at the moment of dying I shall above all doubts, abandon myself."**
> **-Father Teilhard de Chardin**

The world now knows that Teilhard de Chardin had doubts about a fundamental part of his theology, and that perhaps he inserted Christ into the picture solely to please his superiors in the Catholic Church. But that has not stopped a growing number of readers from turning to *The Phenomenon of Man* for the insights it contains.

In Shakespeare's play *Julius Caesar*, Antony tells the crowd come to bury Caesar that "The evil that men do lives after them; the good is oft interred with their bones." Antony didn't really believe this sentiment. He was manipulating the crowd, letting them think first that he agreed with Caesar's assassins that Caesar was an evil man, when all the while he was priming the crowd to side against the assassins once they learned that Caesar had left most of his fortune to the common people of Rome. It was Caesar's good will that lived after him, acts of generosity that were of no benefit to him personally once he was dead.

"Why, man, he doth bestride the narrow world like a Colossus," said Cassius of Caesar. Few men have made so great a mark on human history than Julius Caesar, and what a gulf there was between a Colossus such as he, and the insignificant million or so people who lived in Rome at the time of the Caesars! And yet, while our public institutions have been influenced by men such as Caesar, and our history would have been very different had he not existed, our connection ultimately is not to him, nor other great women or men like him, but to the millions of insignificant people who have passed into historic oblivion over the thousands of years of human history.

Our genetic inheritance derives from these unknown women and men who lived ordinary lives of no particular significance, except for one thing. The majority of them were of service to their fellow man. They raised their families with kindness, they extended help to the needy, they maintained the bonds of community, they advanced human knowledge, and they fought against injustice, cruelty and other human evils. This was their hero project. If they had not given service to humanity, civilization would not have survived, and we would not be here.

Service to humanity is the greatest of callings. It is the essence of Jesus' plea, "Do unto others as you would have them do unto you." It is a basic instinct that has allowed our species to survive and thrive, derived from our infancy and early childhood years when God-like figures provided us love, sustenance, and knowledge. It is this experience which provides all of us with a fundamental understanding of what is moral, and therefore it is whence morality is derived. It is a precept underlying all great religions, but equally, it underlies all great secular philosophies. As is said in the Humanist Manifesto of the American Humanist Association, "Working to benefit society maximizes individual happiness."

Any benefit we can provide society is a positive mark on the human timeline. It is an act of nobility. Few are called to the greatness of a Julius Caesar, but such greatness is not necessary for any of us. Simon Schama, in his book *The History of the Jews*, defined what it was 19th century Jews living in Eastern Europe wanted. They were subject to the most horrific pogroms for over a century, from Czarist police, from anti-Semitic mobs, and ultimately from German nationalists. They were driven from place to place, their homes destroyed, and many of them killed. During this entire period, all that they ever asked for, according to Schama, was "the nobility of an ordinary life."

Your ordinary life is a noble thing. It is of great significance, if only because of the impact you have on people around you. The repercussions of a simple act of charity often ripple out to others in ways you may never know about. Your acts of charity may be performed in a religious context, or in a secular context, with a full recognition in both cases that they reflect a spirit of hope, and a desire for happiness for yourself and others. As was said in the Qur'an:

> **"You cannot attain righteousness until you give to charity from the possessions you love. Whatever you give to charity, God is fully aware thereof."**
> **-The Qur'an, Sura 3:92**

This spirit of giving is reflected in *The Affirmations of Humanism* statement:

> **"We believe in optimism rather than pessimism, hope rather than despair, learning in the place of dogma, truth instead of ignorance, joy rather than guilt or sin, tolerance in the place of fear, love instead of hatred, compassion over selfishness, beauty instead of ugliness, and reason rather than blind faith or irrationality."**

Do you not see the similarities? The Qur'an, as is true for scripture from other religions, is filled with references to compassion and charitable giving to the poor. The Humanist philosophy encourages "compassion over selfishness," and both of these philosophies result in the same outcome – service to humanity. Or perhaps you are fixated on the one difference between the two statements, found at the very end of *The Affirmations* which states that Humanists value "reason rather than blind faith or irrationality."

Certainly more felicitous words could have been chosen here. The overwhelming number of believers in God are rational people. That they have chosen to accept the Illusory God provided them in childhood, and that they, like Mahatma Gandhi, claim that "faith transcends reason," does not make them irrational or delusional. Illusions, after all, have a quality of wish-fulfillment, and all of us survive by living with certain illusions every day – little tricks of the mind that allow us to see the world as we need it to be, rather than how it truly is.

What believers and non-believers also have in common is a Template God resident in our unconscious mind, and two other forms of God from our childhood, which collectively inform us at a deep level that our very existence is owed to people who gave us love, sustenance, and knowledge. We all then underwent the traumas associated with knowledge of death. Some of us took the path of illusion, and acceptance of Illusory God, as a life-long form of comfort and a search for not-death. Others of us followed a path of reason rather than illusion, choosing instead to find purpose in this one life we are allotted on earth. Both paths have the potential to create conflict between believers and non-believers, and history shows how easy it is for one group to oppress another over differences, ignoring the commonalities.

Yet the commonalities do not go away. All of us share the God of our earliest years, who taught us the importance of service to others – of providing love, sustenance, and knowledge to those in need. We each have a longing to overcome death, and to provide some greater, or cosmic meaning to our lives, which can best be achieved by being of service to others. Each of us has the concept of an idealized human being in our minds, which derives from experience with our childhood God.

These are the things all of us seek. We aspire to that which our idealized human being represents. Both faith and reason tell us that our aspirations will never fully be met. We do have flaws.   Some prefer to call these flaws sins, and some prefer to call them imperfections in our human nature. Some prefer to call the idealized human being God, but all of us recognize we are not God, and we are not gods. As was said by one of today's best Catholic thinkers:

> **"Perfection is not what being human is about. The function of being human is to become the best human beings we can be, one insight, one mistake at a time."**
>
> **-Sister Joan Chittister, O.S.B.**

Any secular humanist, and any atheist, would be proud to affirm the truth in Sister Joan's statement. This statement constitutes a bridge between those who believe in God and those who do not, because it reflects the fundamental reality of the human condition, and the potential in all of us to work to improve ourselves and our community.

These bridges of commonality have always been there, but we do not always notice them. Our cultural filters and our Tribal God may make it impossible for us to see them, just as our anger at being ostracized for non-belief in God in any form may cloud our vision to the possibility that there are believers who wish to create a community comprised of both those who do, and those who do not believe in God. One such community are the Unitarian Universalists, whose Seven Principles define a congregation which blends harmoniously both believers and non-believers alike.

The Unitarian Universalists teach us that it is in the interest of humanity, and indeed of all living things on earth, that bridges must be established not just ecumenically among religions, but between religious people and the hundreds of millions of individuals who are secular, and who do not believe in God, or have doubts, and who might not practice a religion. As these bridges are created, it is up to each of us to take the first step in walking over to the other side, to extend a hand in open and sincere friendship. It is up to each of us to see through the fog of anger, to tear down the wall of faith, and to look beyond race and nationality and all such other social divisions, so that we can discover one simple truth.

All of us, *all of us*, once knew God-like figures in our life. The idealized human being of our infancy and childhood is still there, deep inside our unconscious mind, propelling us as if by instinct to seek out those qualities which this God best represented: love, sustenance, and knowledge.

If we are willing to cross over that bridge, to meet "the other," to quietly listen to what they have to say, and if they too are willing to listen to what we have to say, an important discovery can be made: both sides have an aspiration to share as much as possible the love, sustenance, and knowledge that our idealized human being once provided each of us.

All of us once knew heaven. While we learned later that such perfection comes to us only once in our life, we can create a part of heaven on earth in whatever ways are open to us, for the benefit of those around us. What could be of more benefit to humanity and ourselves? Who should be surprised that secular philosophy *and* religious teaching both ask of us to feed the poor, educate the unlearned, heal the sick, visit the imprisoned, and comfort the dying, to create for them as much as possible the conditions of love, sustenance and knowledge that they once knew as infants, and which they, like us, long for as adults?

Even now, the childhood God which was once in your life is speaking to you, in the softest of whispers, reminding you of that which was your birthright. The love, the sustenance, and the knowledge which were given to you freely at your earliest and most vulnerable years were the greatest of gifts, but now that you are an adult, they have become your noblest of obligations. And in the face of the nobility of these obligations to do unto others that which was once done unto you, is it not petty for any of us to hide behind our protective walls of faith, or our fear of "the other," or our tendency to mock that which seems alien to us? In doing so, we ignore the potential benefits that can result when women and men of whatever faith in God, and of no faith in God, combine efforts to fulfill our fundamental human imperative. Can we not see how damaging it is to overlook the reality that *all of us* have a capacity and an obligation to create even a small bit of heaven on earth, with whatever talents and abilities are given to us?

This is your birthright. This is your calling. This is why all religions, and all great secular philosophies, discover that the fundamental moral imperative imposed on us as humans is that we do unto others that which we would have them do unto us. That *which was done unto us, in the beginning.* This, the worthiest of all obligations, is your inheritance, given to you so very long ago by that being you knew as God when you first entered this world.

APPENDIX A

The Seven Principles of Unitarian Universalism

"Unitarian Universalist congregations affirm and promote seven Principles, which we hold as strong values and moral guides. We live out these Principles within a "living tradition" of wisdom and spirituality, drawn from sources as diverse as science, poetry, scripture, and personal experience."

We affirm:

1st Principle: The inherent worth and dignity of every person;

2nd Principle: Justice, equity and compassion in human relations;

3rd Principle: Acceptance of one another and encouragement to spiritual growth in our congregations;

4th Principle: A free and responsible search for truth and meaning;

5th Principle: The right of conscience and the use of the democratic process within our congregations and in society at large;

6th Principle: The goal of world community with peace, liberty, and justice for all;

7th Principle: Respect for the interdependent web of all existence of which we are a part.

APPENDIX B

References and Citations

Akita, Lailah Gifty. *Think Great! Be Great!* Ghana: Amazon Digital Services. 2015.

Ali, Muhammad. *50 Inspirational Pieces of Wisdom from Muhammad Ali.*

Allen, Woody. "Standup Comic CD." 1999.

Aquinas, St. Thomas. *Summa contra Gentiles.*

Asimov, Isaac. "The Foundation Series." *Super Science Stories, John W. Campbell Jr., editor* 1941 - 1949.

Aurelius, Marcus. *Meditations.* Dover Thrift Editions,

Becker, Ernest. *The Denial of Death.* Free Press, Kindle Edition. 1974.

Bloom, Paul. *Just Babies: The Origins of Good and Evil.* Random House. 2014.

Bonhoeffer, Dietrich. *The Cost of Discipleship.* Touchstone. 1995.

Bronte, Emily. *No Coward Soul is Mine, poem.*

Buddha, Guatama. *Buddha Quotes, Brainy Quotes*

Catholic Online. n.d. <http://www.catholic.org/prayers/prayer

Chittister, Sister Joan. *In God's Holy Light: Wisdom from the Desert Monastics.* Franciscan Media. 2015.

Claret, St. Anthony. *The Pains of Hell.* Ignatian Spiritual Exercises and Meditations. 1890.

Confucius. *Analects.* 15:23.

Coolidge, Calvin. *Calvin Coolidge Quotes.* goodreads.com

Coyne, Jerry A. "Einstein's Famous Quote About Science and Religion Didn't Mean What You Were Taught." *The New Republic.* 4 December 2013.

Depression, Bruce Springsteen's. *Pyschcentral.com.* 24 July 2012.

Dostoevsky, Fyodor. *The Brothers Karamazov.*

Dusek, Jeffrey A., et al. "Study of the Therapeutic Effects of Intercessory Prayer on Cardiac Patients." *American Heart Journal*, med.harvard.edu

Edison, Thomas A. "Interview with Thomas Edison." *New York Times Magazine.* 2 October 1910.

Edwards, Jonathan. *Quotes on Hell Fire.* tentmaker.org

Emerson, Ralph Waldo. "Nature." 1836: James Munroe & Co.

Feynman, Richard P. *Richard P. Feynman* BBC Horizon Interview. 1981.

Francis, Pope. *Pope Francis Quotes*, brainyquotes.com

Freud, Sigmund. *The Future of an Illusion*. London, English Version translated by W. D. Robson-Scott: Hogarth Press. 1928.

Frost, Robert. *Birches*. New York City: Henry Holt and Company. 1944.

Furniss, Reverend J. *Christ Triumphant, A Catholic Catechism*, tentmaker.org

Gandhi, Mahatma. *Mahatma Gandhi Quotes*. brainyquotes.com, *Quotes of Mahatma Gandhi*, goodreads.com

Gervais, Ricky. *Piers Morgan on CNN* Piers Morgan. 13 September 2013.

Gray's Anatomy, Plate 237. *studyblue.com*.

Hart, David Bentley. *The Experience of God: Being, Consciousness, Bliss*. Princeton University Press. 2014.

Heschel, Rabbi Abraham Joshua. *Man is Not Alone: A Philosophy of Religion*. Farrar, Strauss, and Giroux. 1976.

Jakes, Reverend T.D. *Quotes from T.D. Jakes*, brainyquotes.com

Jennings, Clayton J. *Heaven. Hell. You, You Tube* Performance by Clayton Jennings. 25 June 2015.

Jung, Carl. *The Archetypes and the Collective Unconscious*. Princeton University Press. 1934.

Kierkegaard, Søren. *The Sickness Unto Death*. English translation Alastair Hannay 1989: Clays Ltd. 1849.

Kirkpatrick, Lee. *Attachment, Evolution, and the Psychology of Religion*. The Guilford Press. 2004

Kowalska, Saint Maria Faustina. *The Visions of Hell, Purgatory and Heaven according to St. Faustina*, cloudoffire.blogspot.com

Lama, Dalai. *Quotes That Will Change Your Life*. 2016, addicted2success.com/quotes

Lewis, C.S. *The Screwtape Letters*. Harper Collins. 1942.

Luther, Martin. *Martin Luther on the Whore of Reason*. joshuasowin.com/archives/

Martel, Yann. *The Life of Pi*. Mariner Books. 2003.

Merton, Thomas. *Thomas Merton Quotes*, goodreads.com/quotes/

Miller, Arthur. *Death of a Salesman*. 1949.

Milton, John. *Paradise Lost*.

Newberg, Andrew, with Mark Waldman. *How God Changes Your Brain*. Ballantine Books. 2009.

Nietzsche, Friedrich. *The Antichrist*. Ernest Schmeitzer. 1878.

Orwell, George. *Down and Out in Paris and London*. Bibliotech Press, Modern Edition. 1933.

Osteen, Joel. *Joel and Victoria's Blog, l*akewoodchurch.com

Pascal, Blaise. *Pensees*. 1670.

Peterson, Cort A. et al. *Oxytocin in Maternal, Sexual, and Social Behaviors*, NY Academy of Sciences. 1992

Pierce, Lara, et al. *"Past Experience Shapes Ongoing Neural Patterns for Language,"* Nature Communications. 2014

Pope, Alexander. *An Essay on Man*. Dover Publications. 1731.

Pratchett, Terry. *Feet of Clay, a Discworld Novel*. Victor Gollancz. 1996. *Reuters*.

Rajaratnam, VS, Talk on Patient With a Severed Corpus Callosum, https://www.youtube.com/watch?v=PFJP tVRlI64&feature=youtu.be 2015. <http://www.reuters.com/article/us-science-psychedelic

Riggedreality, riggedreality.net.

Russell, Bertrand. *Am I an Atheist or an Agnostic?* 1947.

St. Basil the Great. *Prayer of Interdiction to St. Basil the Great,* orthodoxphotos.com/readings/threshold/prayer

Sartre, Jean-Paul, *No Exit*.

Schopenhauer, Arthur. „Religion A Dialogue and Other Essays." Schopenhauer, Arthur. 1855.

Sproul, R.C. *Reason to Believe: A Response to Common Objections to Christianity*. Zondervan. 1982. Also, *The Holiness of God*. Tynsdale House Publishers, 2nd Revised Edition. 2000.

Stormark, Kjell Morten. "Skin Conductance and Heart-Rate Responses as Indices of Covert Face Recognition in Preschool Children." *Infant and Child Development, Wiley InterScience Online* (2004).

Stout, Rex. Interview. Life Magazine. 10 December 1965.

Supplication to Allah, supplication.8m.com

Teilhard de Chardin, Father Pierre, S.J., *Apropos*.org.uk, also *The Heart of the Matter*, Harvest Books, Harcourt Inc. 1976.

Teresa, Mother. *Come Be My Light: The Private Writings of the Saint of Calcutta*. Image Press. 2009.

Tertullian. *De Spectaculis*. c. 200 CE.

Tolstoy, Leo. *War and Peace*.

Warren, Pastor Rick. *Pastor Rick's Daily Hope*.

Wathey, John C. "The Illusion of God's Presence: The Biological Origins of Spiritual Longing." Prometheus Books: Kindle Edition. 2016.

Wright, Frank Lloyd, with Patrick J. Meehan Editor. *Truth Against the World: Frank Lloyd Wright Speaks for an Organic Architecture*. Preservation Press. 1998.

Yogashastra. *Yogashastra (Jain Scripture) Quotes*, shortpoemsandquotes.com/

Zola, Emile. *The Heretic's Handbook of Quotations, Charles Bufe (unconfirmed quote)*. 1992.

APPENDIX C

BIBLIOGRAPHY

Classic Literature on God-Belief and Religion

Books on Atheism

Richard Dawkins, *The God Delusion*

Ludwig Feuerbach, *The Essence of Christianity*

Sigmund Freud, *Death of an Illusion*

Robert G. Ingersoll, *What's God Got to Do with It?*

Critiques of Religion

Steve Allen, *Steve Allen on the Bible, Religion and Morality*

Denis Diderot, *The Skeptic's Walk*

Christopher Hitchens, *God is Not Great*

Friedrich Nietzsche, *The Gay Science*

Thomas Paine, *The Age of Reason*

Bertrand Russell, *Why I Am Not a Christian*

Science in Relation to God, Religion, God-Belief

Pascal Boyer, *Religion Explained*

Daniel Dennett, *Breaking the Spell*

Sam Harris, *Free Will*

Lawrence Krause, *A Universe from Nothing*

Andrew Newberg and Mark Waldman, *How God Changes Your Brain*

John Wathey, *The Illusion of God's Presence*

Deconversion Accounts

Greta Christina, *Why Are You Atheists So Angry?*

Michael Vito Tosto, *Portrait of an An Infidel*

In Defense of God and Religion

David Bentley Hart, *The Experience of God*

Alistair McGrath, *Why God Won't Go Away*

Ravi Zacharias, *Can Man Live Without God*

APPENDIX D

ABOUT THE AUTHOR

GARRETT GLASS is the author of the acclaimed Jehoshua series of historical fiction novels, based on the first century of Christianity. Book One of the Series, *Jehoshua: Signs and Wonders*, covers the first 25 years of Christianity. Glass, according to Kirkus Reviews, "skillfully fleshes out the characters and illustrates the theological issues without confusing the reader...well researched and highly informative." *Jehoshua: Signs and Wonders,* was the first self-published book to be included in the prestigious Patheos Book Club sponsored by the editors of patheos.com, the world's most-visited website devoted to religious matters. Book Two of the series, *Jehoshua: Conflagration*, details the next 25 years of Christianity, up to the destruction of Jerusalem by the Romans. Book Three is currently being written, and chronicles the establishment of formal churches among the Christian communities, a development that solidified control over the Christian movement by men, and the forcible expulsion of women from an active role in the ministry.